Lecture Notes in Mathematics 1623

Editors:
A. Dold, Heidelberg
F. Takens, Groningen

Springer
Berlin
Heidelberg
New York
Barcelona
Budapest
Hong Kong
London
Milan
Paris
Santa Clara
Singapore
Tokyo

H. P. Yap

Total Colourings
of Graphs

 Springer

Author

Hian-Poh Yap
National University of Singapore
Department of Mathematics
Lower Kent Ridge Road
Singapore 0511, Republic of Singapore

Cataloging-in-Publication Data applied for

Die Deutsche Bibliothek - CIP-Einheitsaufnahme

Yap, Hian Poh:
Total colourings of graphs / Hian Poh Yap. - Berlin ;
Heidelberg ; New York ; Barcelona ; Budapest ; Hong Kong ;
London ; Milan ; Paris ; Tokyo : Springer, 1996
  (Lecture notes in mathematics ; 1623)
  ISBN 3-540-60717-X
NE: GT

Mathematics Subject Classification (1991): 05C15, 05C35, 05C75

ISBN 3-540-60717-X Springer-Verlag Berlin Heidelberg New York

© Springer-Verlag Berlin Heidelberg 1996
Printed in Germany

Typesetting: Camera-ready T$_E$X output by the author
SPIN: 10479714   46/3142-543210 - Printed on acid-free paper

Preface

I started writing this book in 1990 and completed the first draft in October 1991. It then took me another one and a half years (June 1992 to December 1993) to revise the first draft. My objective in writing this book is to give an up-to-date account of total colourings of graphs which can be used as a graph theory course/seminar materials for advanced undergraduate and graduate students and as a reference for researchers. To achieve the objectives, easy-to-read, detailed proofs of almost all of the theorems presented in this book, and numerous examples and exercises are provided here. Many open problems are also mentioned. I hope that through this rapid introduction I shall be able to bring the readers to the frontier of this currently very active field in graph theory.

After the first draft of this manuscript was completed, I used it as lecturing material in my graph theory course offered to the advanced undergraduate students of the National University of Singapore (NUS). I thank my students for their patience in attending my lectures and for giving me their valuable feedback.

I would like to thank the NUS for granting me a 10-month (1 July 1991 to 30 April 1992) sabbatical leave so that I could concentrate on writing and revising the manuscript. I would also like to thank the NUS for granting me conference leaves (June 1993 and June 1995) and to the Japan Society for the Promotion of Science for sponsoring my visit to three universities in Japan (12 April to May 1, 1993) so that I could have direct discussions with many graph theorists working on total colourings, and at the same time popularize this subject by giving many survey talks on various topics covered in this book. I had altogether given at least forty survey talks on total colourings of graphs to the following 18 institutions in Taiwan, People's Republic of China, USA, Japan and Singapore during the past three and half years: Academia Sinica (Taipei) and National Chiao Tung University, Hsingchu, Taiwan (1 - 30 November, 1991); Beijing Institute of Technology, Tsinghua University, Institute of Applied Mathematics, Academia Sinica (Beijing), Lanzhou Railway Institute, Lanzhou University, Shaanxi Normal University (Xian), Zhengzhou University and Institute of Systems Science, Academia Sinica (Beijing) (8 December 1991 to 6 January 1992); West Virginia University, USA (14 January to 8 March, 1992); Spring School and International Conference on Combinatorics held at Lushan Mountain and Hungshan Mountain, People's Republic of China (10 April to 30 April, 1992); Science University of Tokyo, Ibaraki University and Keio University, Japan (12 April to May 1, 1993); Inner Mongolia University and Taiyuan Institute of Machinery, People's Republic of China (June, 1993); Spring School and International Conference on Combinatorics held at Hefei, People's Republic of China (May 22 to June 3, 1995). It is also a great pleasure for me to acknowledge the helpful comments and suggestions received from many friends who hosted my visits : Dr. Bor-Liang Chen, Professors A. J. W. Hilton, Zhang Zhongfu, Yoshimi Egawa, Mikio Kano, Hikoe Enomoto, Ku Tung-Hsin and Li Jiong-Sheng. I would like to express my deepest gratitude to Dr. Hugh R. Hind for carefully reading the first draft and making many valuable comments and suggestions. I am also thankful to Dr. Abdón Sánchez-Arroyo, Mr. Zhang Yi and Professor A. D. Keedwell in proofreading the second draft, to Mr. Liu Qizhang for using computers to draw the figures, to Professors J. C. Bermond, O. V. Borodin, A. V. Kostochka and C. J. H. McDiarmid for sending preprints and reprints of their papers to me, and to Miss D. Shanthi for typing this book in PCTEX.

Finally, a few words on the reference system and the exercises of this book. When a research paper by XX is referenced in the text as XX [93], it denotes that the paper by XX in the List of References was published in 1993. When a paper is referenced as YY [-a], it is unpublished and the ordering a, b, c, ... reflects the ordering of the unpublished papers of YY in the List of References. When an exercise is marked with a minus sign or a plus sign, it means that the exercise is easy or hard/time-consuming respectively; and if it is marked with a star, it means that it is an open problem or a conjecture.

H. P. Yap
November 8, 1995

CONTENTS

Chapter 5: GRAPHS OF HIGH DEGREE

Chapter 6: CLASSIFICATION OF TYPE 1 and TYPE 2 GRAPHS

Chapter 7: TOTAL CHROMATIC NUMBER OF PLANAR GRAPHS

Chapter 8: SOME UPPER BOUNDS FOR THE TOTAL CHROMATIC NUMBER OF GRAPHS

Chapter 9: CONCLUDING REMARKS

CHAPTER 1

BASIC TERMINOLOGY AND
INTRODUCTION

§1. Basic terminology

In this section we define some basic terms that will be used in this book. Other terms will be defined when they are needed.

Unless stated otherwise, all graphs dealt in this book are finite, undirected, simple and loopless. Let $G = (V, E)$ be a graph, where $V = V(G)$ is its vertex set and $E = E(G)$ is its edge set. For a graph G, we denote $VE(G) = V(G) \cup E(G)$. The <u>order</u> of G is the cardinality $|V|$ of V and is denoted by $|G|$ or $v(G)$. The <u>size</u> of G is the cardinality $|E|$ of E and is denoted by $e(G)$. Two vertices u and v of G are said to be <u>adjacent</u> if $uv \in E$. If $e = uv \in E$, then we say that u and v are the <u>end-vertices</u> of e and that the edge e is <u>incident</u> with u and v. Two edges e and e' of G are said to be <u>adjacent</u> if they have one common end-vertex. If $uv \in E$, then we say that v is a <u>neighbour</u> of u. The set of all neighbours of u is called the <u>neighbourhood</u> of u and is denoted by $N_G(u)$ or simply by $N(u)$ if there is no danger of confusion. The <u>degree</u> (<u>valency</u>) of a vertex u is $|N(u)|$ and is denoted by $d_G(u)$ or simply by $d(u)$. The maximum (resp. minimum) of the vertex degrees of G is called the <u>maximum</u> (resp. <u>minimum</u>) <u>degree</u> of G and is denoted by $\Delta(G)$ (resp. $\delta(G)$).

A graph H is said to be a <u>subgraph</u> of a graph G if $V(H) \subseteq V(G)$ and $E(H) \subseteq E(G)$. We write $H \subseteq G$ if H is a subgraph of G. A subgraph H of G such that whenever $u, v \in V(H)$ are adajcent in G then they are also adjacent in H is called an <u>induced subgraph</u> of G. We write $H \leq G$ if H is an induced subgraph of G. An induced subgraph of G having vertex set V' is denoted by $G[V']$. The subgraph of G induced by $V(G) \backslash \{v_1, ..., v_k\}$, where $\{v_1, ..., v_k\} \subseteq V(G)$, is written as $G - \{v_1, ..., v_k\}$ or $G - v_1 - ... - v_k$ when k is small. The subgraph of G having edge set $E' \subseteq E(G)$ and vertex set the set of end-vertices of all edges in E' is denoted by $G[E']$. The subgraph of G having vertex set $V(G)$ and edge set $E(G) \setminus E'$, where $E' \subseteq E(G)$, is denoted by $G - E'$, and in particular, if E' consists of only a small number of edges $e_1, ..., e_k$, then this subgraph is denoted by $G - e_1 - ... - e_k$. The complement $\bar{G}$ of G is the graph having vertex set $V(G)$ and edge set $\{uv | u, v \in V(G), uv \notin E(G)\}$. If $E' \subseteq E(\bar{G})$, then $G + E'$ is the graph having vertex set $V(G)$ and edge set $E(G) \cup E'$.

If all the vertices of a graph G have the same degree d, then we say that G is regular of degree d, or G is a d-regular graph. The degree of a regular graph G is written as $\deg(G)$. A regular graph of degree 3 is called a cubic graph. If G is regular graph of order n such that $\deg(G) = 0$ (resp. $n - 1$), then G is called a null graph (resp. complete graph) and is denoted by O_n (resp. K_n).

A set of vertices S of a graph G is said to be independent if any two vertices u and v in S are not adjacent in G. The maximum cardinality of an independent set of vertices of G is called the vertex-independence number and is denoted by $\alpha(G)$. Analogously, a set of edges E' of a graph G is said to be independent if any two edges e and e' in E' are not adjacent in G. The maximum cardinality of an independent set of edges of G is called the edge-independence number of G is denoted by $\alpha'(G)$. An independent set of edges of G is also called a matching in G. A matching in G that includes (saturates) every vertex of G is called a perfect matching or a 1-factor of G. A matching of G that saturates every vertex, except one, of G is called a near-perfect matching of G. Thus G has a perfect matching only if $|G|$ is even and G has a near perfect matching only if $|G|$ is odd.

If the vertex set of a graph G can be partitioned into r independent sets $V_1, ..., V_r$ then G is called an r-partite graph (when $r = 2$, G is called a bipartite graph having bipartition (V_1, V_2)). Moreover, if every vertex of V_i is joined to every vertex of V_j, $j \neq i$, then G is called a complete r-partite graph. We denote a complete r-partite graph by $O_{p_1} + O_{p_2} + ... + O_{p_r}$ where $p_i = |V_i|, i = 1, 2, ..., r$. If $G = O_{p_1} + O_{p_2} + ... + O_{p_r}$ and $p_1 = p_2 = ... = p_r$, then we call G a balanced complete r-partite graph and we denote such a graph by $K(r, n)$, where $n = p_1 = p_2 = ... = p_r$. A spanning subgraph of a balanced complete r-partite graph is called a balanced r-partite graph. A bipartite graph having bipartition (V_1, V_2) such that $|V_1| = m$ and $|V_2| = n$ is denoted by $K_{m,n}$. The graph $K_{1,r}$ is called a star and is denoted by S_r.

A cycle of length n is denoted by C_n and a (shortest) path of length n is denoted by P_n. If G has a cycle C that includes every vertex of G, then C is called a Hamilton cycle of G and G is said to be hamiltonian. If G has a path P that includes every vertex of G, then P is called a Hamilton path of G. Two graphs G and H are said to be disjoint if they have no vertex in common. The join $G + H$ of two disjoint graphs G and H is the graph having vertex set $V(G) \cup V(H)$ and edge set $E(G) \cup E(H) \cup \{xy | x \in V(G), y \in V(H)\}$. The union $G \cup H$ of two disjoint graphs G and H is the graph having vertex set $V(G) \cup V(H)$ and edge set $E(G) \cup E(H)$.

Suppose G and H are two disjoint graphs. If there exists an injection $\phi : V(G) \to V(H)$ such that $\phi(x)\phi(h) \in E(H)$ if $xy \in E(G)$, then we say that ϕ is an embedding of G in H. If such an embedding exists, then we say that G is embeddable in H.

A multigraph permits more than one edge joining two of its vertices. In a multigraph G the number of edges joining two vertices x and y of G is called the multiplicity of xy and is denoted by $\mu(x, y)$. The multiplicity $\mu(G)$ of a multigraph G is max $\mu(x, y)$ taken over all pairs of adjacent vertices x and y of G.

A (proper) vertex-colouring (resp. edge-colouring) of a graph G is a mapping φ from $V(G)$ (resp. $E(G)$) to a set $\mathcal{C}$ such that no adjacent vertices (resp. edges) of G have the same image. If $\varphi : V(G) \to \mathcal{C}$ (resp. $\varphi : E(G) \to \mathcal{C}$) is a vertex-colouring (resp. edge-colouring) of G and $|\mathcal{C}| = k$, a positive integer, then we say that G is k-colourable (resp. k-edge-colourable), and φ is called a k-colouring or k-vertex-colouring (resp. k-edge-colouring) of G. The minimum cardinality of $\mathcal{C}$ for which there exists a vertex-colouring $\varphi : V(G) \to \mathcal{C}$ (resp. an edge-colouring $\varphi : E(G) \to \mathcal{C}$) is called the chromatic number (resp. chromatic index) of G, and is denoted by $\chi(G)$ (resp. $\chi'(G)$). If φ is a k-vertex-colouring (resp. k-edge-colouring) of G, then φ yields a partition of $V(G)$ (resp. $E(G)$) into independent sets $V_1, ..., V_k$ (resp. $E_1, ..., E_k$). These independent sets $V_1, ..., V_k$ of vertices (resp. $E_1, ..., E_k$ of edges) of G are called the colour classes of φ.

Suppose $Z_n = \{0, 1, ..., n-1\}$ is the group under addition modulo n. Let $S \subseteq Z_n$ be such that $0 \notin S$ and if $s \in S$, then $-s \in S$. The circulant graph $G(Z_n, S)$ is the simple graph having vertex set Z_n and edge set $\{\{g, h\} \mid h - g \in S\}$. The set S is called the symbol of the circulant graph $G(Z_n, S)$.

A planar graph G is outerplanar if it can be drawn on a plane in such a way that G has no crossings and that all its vertices lie on the boundary of the same face.

§2. Total-colouring of a graph - introduction

A total-colouring π of a graph G is a mapping from $VE(G)$ to a set $\mathcal{C}$ satisfying :

(i) no two adjacent vertices or edges of G have the same image; and

(ii) the image of each vertex of G is distinct from the images of its incident edges.

If $\pi : VE(G) \rightarrow \mathcal{C}$ is a total-colouring of G and $|\mathcal{C}| = k$, a positive integer, then we say that G is k-total-colourable. The minimum cardinality of $\mathcal{C}$ for which there exists a total-colouring $\pi : VE(G) \rightarrow \mathcal{C}$ is called the total chromatic number of G, and is denoted by $\chi_T(G)$. Thus if π is a total-colouring of G, then $\pi|_{V(G)}$, the restriction of π on $V(G)$, is a vertex-colouring of G. Similarly, $\pi|_{E(G)}$ is an edge-colouring of G. From this, it follows that a total-colouring π of G yields a partition of $VE(G)$ into independent sets $V_1 \cup E_1$, $V_2 \cup E_2$,..., where $V_1, V_2,...$ are independent sets of vertices of G and $E_1, E_2,...$ are independent sets of edges of G, and that no vertex in V_i is incident with any edge in E_i. These independent sets $V_1 \cup E_1$, $V_2 \cup E_2$,... are called the colour classes of π. Conversely, any partition of $VE(G)$ into independent sets $V_1 \cup E_1, V_2 \cup E_2, ..., V_k \cup E_k$ gives rise to a k-total-colouring of G.

Undoubtedly, vertex-colourings and edge-colourings are among the main streams in Graph Theory. These two topics both have long histories. They are very important, difficult, and they have many real-life applications to storage problem, timetabling problem, electrical networks, production scheduling, and designs for experiments etc. Since a total-colouring of G is a vertex-colouring and at the same time an edge-colouring of G, the degree of difficulty of this subject is obvious and its importance is anticipated. Moreover, it is not surprising that very soon some nontrivial and important applications of total-colourings of graphs will be found.

The notion of a total-colouring of a graph was introduced and studied by Behzad and independently Vizing around the year 1965. Clearly, for any graph G, $\Delta(G)+1 \leq \chi_T(G)$, where $\Delta(G)$ is the maximum degree of G. The following conjecture was posed independently by Behzad and Vizing in 1965.

Total Colouring Conjecture (TCC) : For any graph G,

$$\chi_T(G) \leq \Delta(G) + 2.$$

(In fact, Vizing posed a more general conjecture which says that for any multigraph

G, $\chi_T(G) \leq \Delta(G) + \mu(G) + 1$.)

The TCC was proved true for a few classes of graphs in the 1970s. Only very recently, some new techniques have been introduced and used to prove that the TCC holds for some more classes of graphs, especially graphs having high maximum degree. In this book, we shall give an up-to-date account on results obtained in this area.

In chapter 2 some basic results on total-colourings of graphs are given. These basic results will be used very often throughout this book. Amongst these basic results are : (i) a powerful lemma which says that if a graph G contains an independent set of vertices S such that $|S| \geq |G| - \Delta(G) - 1$, then $\chi_T(G) \leq \Delta(G) + 2$ (This lemma will be used in Chapter 3 to show that complete r-partite graphs satisfy the TCC and it will also be used in Chapter 6 to show that graphs of high maximum degree satisfy the TCC); (ii) a useful theorem saying that if G is a graph of order $2n$ and $\chi_T(G) = t + 1$, then $e(\bar{G}) + \alpha'(\bar{G}) \geq n(2n - t)$. (This theorem will be used to show that some complete r-partite graphs G of even order has $\chi_T(G) = \Delta(G) + 2$ in Chapter 3 and many other results in Chapter 6.)

In Chapter 3 the exact value of $\chi_T(G)$ for $G = K_n$ and $G = K_{m,n}$ are determined. The main objectives of this chapter are: (i) to prove that the complete r-partite graphs satisfy the TCC; (ii) to prove that every complete r-partite graph of odd order has total chromatic number $\Delta(G) + 1$; (iii) to give a complete classification of balanced complete r-partite graphs according to their total chromatic numbers.

In Chapter 4 different proof techniques are used to show that the TCC holds for graphs G having $\Delta(G) = 3$ and $\Delta(G) = 4$.

In Chapter 5 it is proved that the TCC holds for graphs G having $\Delta(G) \geq |G| - 5$ and for graphs G having $\Delta(G) \geq \frac{3}{4}|G|$.

In Chapter 6 the exact value of $\chi_T(G)$ for graphs G having $\Delta(G) \geq |G| - 2$ are determined. The exact value of $\chi_T(G)$, where $G = K_{n,n} - E(J)$, $J \subseteq K_{n,n}$ and $\Delta(G) = n$ is given without proof. Some partial results on the total chromatic number of graphs having $\Delta(G) = |G| - 3$ are presented. Finally a complete classification (according to their total chromatic numbers) of regular graphs G whose complement $\bar{G}$ is bipartite is also stated without proof.

Chapter 7 is devoted to the study of total chromatic number of planar graphs. In this chapter we prove that the TCC holds for planar graphs G having $\Delta(G) \geq 8$

and we also prove that for planar graphs G having $\Delta(G) \geq 14$, $\chi_T(G) = \Delta(G) + 1$.

In Chapter 8 the following upper bounds for $\chi_T(G)$ are presented: (i) $\chi_T(G) \leq \chi'(G) + \lfloor \frac{1}{3}\chi(G) \rfloor + 2$; (ii) $\chi_T(G) \leq \chi'(G) + 2\lceil \sqrt{\chi(G)} \rceil$; (iii) $\chi_T(G) \leq \chi'(G) + k$ where k is the smallest positive integer such that $k! \geq v(G)$; and (iv) $\chi_T(G) \leq \Delta(G) + 2\lceil \frac{v(G)}{\Delta(G)} \rceil + 1$. Again, the techniques used to prove these results are totally different.

In Chapter 9 we mention some other results on/or related to total-colourings of graphs which have not been discussed in the previous sections. Probably they too will have some impact on future research on total-colourings.

Exercise 1

1. Prove that for any integer $n \geq 3$,

$$\chi_T(C_n) = \begin{cases} 3 & \text{if } n \equiv 0 \; (mod \; 3) \\ 4 & \text{otherwise.} \end{cases}$$

2. Show that $\chi_T(G) = \Delta(G) + 1$ where G is a tree of order at least 3.

3. Let G be a graph having $\chi_T(G) = t$. Suppose for any t-total-colouring π of G and for any colour class $V_i \cup E_i$ ($V_i \subseteq V(G)$ and $E_i \subseteq E(G)$), we have $|E_i| \geq 2$. Prove that for any edge e of G, $\chi_T(G - e) = t$.

4.* Let G be a graph having $\chi_T(G) = t$. Suppose G has a t-total-colouring π such that π has a colour class $V_m \cup E_m$ for which $|E_m| \geq 3$ is minimum among all possible colour classes of any t-total-colouring of G. Clearly if $e' \in E_m$, then by Exercise 1(2), $G' = G - e'$ has $\chi_T(G') = t$ and G' has a t-total-colouring φ such that φ has a colour class $V_m' \cup E_m'$ for which $|E'| \leq |E_m| - 1$. Is it true that $|E_m| - 1$ is the minimum cardinality of E_i' for any colour class $V_i' \cup E_i'$ of any t-total-colouring of G'?

5. Prove that for any graph $G \neq K_2$ and any edge e of G,

$$\chi_T(G - e) \geq \chi_T(G) - 1.$$

(Behzad [71b])

CHAPTER 2

SOME BASIC RESULTS

Similar to the study of vertex-colourings and edge-colourings of graphs, in the study of total-colourings of a graph G, we shall always assume that G is connected. In this chapter we present some basic results which will be used very often in this book.

The following lemma is often used either implicitly or explicitly. This lemma requires no proof.

Lemma 2.1 For any subgraph H of a graph G, $\chi_T(H) \leq \chi_T(G)$.

The following theorem is due to König [36; Chapter 11].

Theorem 2.2 Every graph (resp. multigraph) G having maximum degree k can be embedded into a k-regular graph (resp. multigraph).

Proof. We take two copies of G and join two corresponding vertices v and v' by an edge if $d(v) < k$. (If G is a multigraph we join the two corresponding vertices by $k - d(v)$ edges and we straightaway obtain a k-regular multigraph H in which G is embedded.) Now the minimum degree of this new graph G_1 is $\delta(G) + 1$ and $\Delta(G_1) = \Delta(G)$. We continue the same process if G_1 is not regular and eventually (after at most k steps) we obtain a k-regular graph H in which G is embedded. //

From Lemma 2.1 and Theorem 2.2 we can deduce the following theorem, which is useful in proving that the TCC holds for graphs having low maximum degree.

Theorem 2.3 (Behzad [71b]) If the TCC holds for all Δ-regular graphs, then it holds for any graph G having maximum degree Δ.

Proofs of the following theorem can be found in many books on graph theory, for instance, in Yap [86].

Theorem 2.4 (Vizing [64]) If G is a multigraph having maximum degree Δ and maximum multiplicity μ, then

$$\chi'(G) \le \Delta + \mu.$$

In particular, if G is a simple graph having maximum degree Δ, then $\chi'(G) = \Delta$ or $\chi'(G) = \Delta + 1$.

A graph G is said to be <u>Class 1</u> if $\chi'(G) = \Delta(G)$ and <u>Class 2</u> if $\chi'(G) = \Delta(G)+1$. If the TCC holds for a certain class of graphs G, then we say that G is <u>Type 1</u> if $\chi_T(G) = \Delta(G) + 1$ and is <u>Type 2</u> if $\chi_T(G) = \Delta(G) + 2$. This definition is analogous to the above definition of Class 1 and Class 2 graphs in edge-colourings of a graph.

If v is a vertex of degree $\Delta(G)$ in G, then v is called a <u>major vertex</u> of G, otherwise a <u>minor vertex</u> of G. Suppose G has maximum degree Δ. The <u>core</u> of G is the subgraph of G induced by the major vertices of G and is denoted by G_Δ.

The following lemma follows immediately from some results of Vizing (Theorem 3.3 and Corollary 3.6 in Yap [86]).

Lemma 2.5 Suppose G is a graph having maximum degree Δ. If G_Δ is a forest, then G is Class 1.

The new technique used in the proof of the following lemma was introduced independently and almost at the same time (around 1986) by A. G. Chetwynd and A. J. W. Hilton, as well as by H. P. Yap, Wang Jian-Fang and Zhang Zhongfu.

Lemma 2.6 (Yap, Wang and Zhang [89]) Let G be a graph of order n and let $\Delta = \Delta(G)$. If G contains an independent set S of vertices, where $|S| \ge n - \Delta - 1$, then

$$\chi_T(G) \le \Delta + 2.$$

Proof. Let M be a maximal matching in $G - S$ and let G^* be a graph obtained by adjoining a new vertex $v^* \notin V(G)$ to $G - M$ and adding an edge joining v^* to each vertex in $G - M - S$. Now $\Delta + 1 \ge \Delta(G^*) \ge \Delta$. If $\Delta(G^*) = \Delta$, then by Theorem 2.4, $\chi'(G^*) \le \Delta + 1$. On the other hand, if $\Delta(G^*) = \Delta + 1$, then the core of G^* is a forest and thus by Lemma 2.5, $\chi'(G^*) = \Delta + 1$. Let φ be an edge-colouring of G^*

using colours $1, 2, ..., \Delta + 1$. We now modify φ to a total-colouring π of G by setting:

$$\pi(v) = \varphi(v^*v) \qquad \text{for each } v \in V(G - S),$$

$$\pi(v) = \Delta + 2 \qquad \text{for each } v \in S,$$

$$\pi(e) = \varphi(e) \qquad \text{for each } e \in E(G - M), \text{ and}$$

$$\pi(e) = \Delta + 2 \qquad \text{for each } e \in M. \qquad //$$

The following is a generalization of a result of Hilton [89/90]. We shall see in the subsequent chapters that this generalized result unifies several previous results and proof techniques of J. C. Bermond, B. L. Chen and H. L. Fu, A. J. W. Hilton, as well as K. H. Chew and H. P. Yap.

Theorem 2.7 (Hilton [89/90]; Yap [95]) Suppose G is a graph of order $2n$ and $\chi_T(G) = t + 1$. Then

$$e(\bar{G}) + \alpha'(\bar{G}) \geq n(2n - t).$$

Proof. Let $m = \alpha'(\bar{G})$. Suppose φ is a $(t + 1)$-total-colouring of G. Let

$$\varphi(V(G)) = \{\varphi(v) | v \in V(G)\} = \{c_1, c_2, ..., c_k\} = C.$$

It is clear that each colour class V_i of vertices (of φ) forms a clique in $\bar{G}$. If $|V_i| = 2s$ or $2s + 1$, we add s independent edges of $\bar{G}$ in $G[V_i]$. Let E' be the set of edges added in $G[V_1] \cup ... \cup G[V_k]$. Then $p = |E'| \leq m$.

Next, let H be the graph obtained from $G + E'$ by adjoining a new vertex v^* and adding an edge joining v^* to each vertex in $V(G) \setminus V(E')$. Then $\varphi|_{E(G)}$, the restriction of φ to $E(G)$, can be extended to a $(t + 1)$-edge-colouring of H by setting $\varphi(e') = \varphi(w)$ if $e' \in E'$ and w is an end-vertex of e', and $\varphi(v^*u) = \varphi(u)$ for any $u \in V(G) \setminus V(E')$. Observe that

$$e(H) = e(G) + (2n - 2p) + p = (n(2n - 1) - e(\bar{G})) + 2n - p$$

$$= n(2n + 1) - (e(\bar{G}) + p).$$

Since each edge-colour class E_i of φ contains at most $\lfloor \frac{v(H)}{2} \rfloor = n$ edges and $\chi'(H) \leq (t + 1)$, we have

$$n(2n + 1) - (e(\bar{G}) + p) \leq (t + 1)n.$$

Consequently,

$$e(\bar{G}) + \alpha'(\bar{G}) \geq e(\bar{G}) + p \geq n(2n - t). \qquad //$$

From the last line of the proof of Theorem 2.7, we know that if $e(\bar{G}) + \alpha'(\bar{G}) = n(2n - t)$, then $p = |E'| = \alpha'(\bar{G}) = m$. Hence if $\bar{G}$ does not induce K_4, then $|V_i| = 2$ or 3 for exactly m colour classes V_i. Thus we have the following corollary.

Corollary 2.8 Let G be a graph of order $2n$. If G satisfies $e(\bar{G}) + \alpha'(\bar{G}) = n(2n - \Delta(G))$, $\bar{G}$ does not induce K_4, and G is Type 1, then for any $(\Delta(G) + 1)$-total-colouring of G, there are $m = \alpha'(\bar{G})$ pairs of pairwise nonadjacent vertices $\{x_i, y_i\}$, $i = 1, 2, ..., m$ receiving m distinct colours. In particular, if G is regular, $\bar{G}$ does not induce K_4, and is Type 1, then $\alpha'(\bar{G}) = n$ and thus $\bar{G}$ contains a 1-factor $x_i y_i$, $i = 1, 2, ..., n$ such that $\{x_i, y_i\}$, $i = 1, 2, ..., n$ receive n distinct colours in any $(\Delta(G) + 1)$-total-colouring of G.

Remarks. Suppose G is a graph of order $2n$ and G is Type 1. Then $e(\bar{G}) + \alpha'(\bar{G}) \geq n(2n - \Delta(G))$. However this necessary condition in general is not a sufficient condition for G to be Type 1. Lemma 6.4 provides such examples.

The maximum cardinality of an independent set of elements in $VE(G)$ is called the <u>total independence number</u> of G and is denoted by $\alpha_T(G)$. Suppose S is a maximum independent set of vertices in G and M is a perfect matching in $G - S$ or a near perfect matching of $G - S$, then clearly

$$\alpha_T(G) = |S| + \left\lfloor \frac{|G| - |S|}{2} \right\rfloor.$$

We observe that each element in an independent set $I \subseteq VE(G)$ consists of either a vertex or an edge, and if an edge is exchanged for two independent vertices, then the size of I increases. This simple observation can be stated as a lemma.

Lemma 2.9 If G contains a maximum independent set of vertices S and $G - S$ contains an independent set of edges E' such that $|E'| = \left\lfloor \frac{|G| - |S|}{2} \right\rfloor$, then

$$\alpha_T(G) = |S| + \left\lfloor \frac{|G| - |S|}{2} \right\rfloor.$$

Hence, in general, $\alpha_T(G) \leq \alpha(G) + \left\lfloor \frac{|G| - \alpha(G)}{2} \right\rfloor$.

Let $\Delta = \Delta(G)$. Suppose $\chi_T(G) = \Delta + 1$ and π is a $(\Delta + 1)$-total-colouring of G using colours $c_1, ..., c_s$, where $s = \Delta + 1$. Let n_i be the number of vertices of G coloured c_i and r be the number of n_i's such that $n_i \equiv v(G) \pmod 2$. We shall now show that $\text{def}(G) \geq s - r$, where the <u>deficiency</u> $\text{def}(G)$ of G is defined by

$$\text{def}(G) = \sum_{v \in V(G)} (\Delta - d(v)).$$

Since r of the numbers $n_1, n_2, ..., n_s$ are of the same parity as $v(G)$, the remaining $s - r$ numbers n_i are of opposite parity as $v(G)$. Since each of these $s - r$ colours c_i is absent at least one vertex, $\text{def}(G) \geq s - r$.

Suppose π is an s-vertex-colouring (not necessarily an s-total-colouring) of G, where $s = \Delta(G) + 1$ and the numbers n_i, $i = 1, ..., s$ and r are given as above. If $\text{def}(G) \geq s - r$, then π is called a <u>conformable colouring</u> of G. If G has a conformable vertex-colouring, then G is said to be <u>conformable</u>. Note that if G is conformable, then it has a spanning vertex-induced subgraph of the form $O_{n_1} \cup O_{n_2} \cup ... \cup O_{n_s}$. Note also that an s-vertex-colouring (respectively s-edge-colouring, s-total-colouring) need not to use up all the s colours. The relationship between Type 1 graphs and conformable graphs proved above can now be stated as follows:

Lemma 2.10 (Chetwynd and Hilton [88]) Every Type 1 graph is conformable.

The following result follows immediately from the definition of conformability and Lemma 2.10.

Corollary 2.11 Let G be a regular graph. Suppose G is conformable and π is a conformable vertex-colouring of G using colours $c_1, ..., c_s$, where $s = \Delta(G) + 1$. Let n_i be the number of vertices of G coloured c_i. Then for each $i = 1, 2, ..., s$, n_i and $v(G)$ are of the same parity.

The following is a conformable graph which is not Type 1. A conformable vertex-colouring using four colours c_1, c_2, c_3 and c_4 is depicted. It is not difficult to show that $\chi_T(G) = 5$ (see Exercise 2(1)).

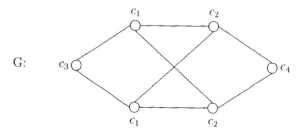

Figure 2.1

In chapter 3, we shall prove that for any $n \geq 1$, $K_{n,n}$ is Type 2. Clearly for each even $n \geq 2$, $K_{n,n}$ is conformable. Hence, in general, conformability is only a necessary condition but not a sufficient condition for a graph G to be Type 1.

Exercise 2

1. Let G be a graph obtained from $K_{3,3}$ by deleting an edge. Prove that $\chi_T(G) = 5$.

2. Let G be a complete bipartite graph having bipartition (X, Y), where $|X| = |Y| = n \geq 3$. Prove or disprove that for any total-colouring φ of G using $n + 2$ colours, X and Y must form two colour classes of φ.

3.⁻ Let G be a graph having exactly one vertex of degree $v(G) - 1$. Prove that $\chi_T(G) = v(G)$.

4.* Let G be a graph having exactly one major vertex. Is it true in general that $\chi_T(G) = \Delta(G) + 1$? (This problem is contributed by Zhang Zhongfu. It has been proved affirmative for G having maximum degree 2 and 3.)

5. Suppose G is a $(2n - 3)$-regular graph of order $2n$. If G is Type 1, prove that $\bar{G}$ is the union of even cycles. (Hint: Apply Theorem 2.7; Wang [90])

6. A graph G is called a <u>polar graph</u> and is denoted by $G(K_n, O_r)$ if $V(G)$ is the disjoint union of V_1 and V_2 such that $G[V_1] = K_n$ and $G[V_2] = O_r$. Prove that every polar graph satisfies the TCC. (Hint: Apply Lemma 2.6; Chen, Fu and Ko [-a])

7.⁻ Suppose G is a Type 1 balanced bipartite graph having bipartition (X, Y). Let $\Delta = \Delta(G)$ and $s = \Delta + 1$. Suppose π is an s-total-colouring of G using colours $c_1, ..., c_s$. Let m_i (resp. n_i) be the number of vertices in X (resp. Y) coloured with colour c_i. Prove that

$$\text{def}(G) \geq \sum_{i=1}^{s} |m_i - n_i|.$$

(Chetwynd and Hilton [88])

8.* <u>Hilton's Conjecture</u>: Let G be a graph having $\Delta(G) \geq \frac{1}{2}(v(G) + 1)$. Then G is not Type 1 if and only if either G contains a non-conformable subgraph H with $\Delta(H) = \Delta(G)$, or $\Delta(G)$ is even and G contains a subgraph H obtained by inserting a new vertex into an edge of $K_{\Delta(G)+1}$.

9.⁻ Let G be a graph and let $\Delta = \Delta(G)$. If W is a subset of $V(G)$, let $V_{<\Delta}(W)$ denote the set of vertices in W which have degree less than Δ in the graph G. Following the notation given in Exercise 2(7) above, we call an s-vertex-colouring of a balanced bipartite graph G with bipartition (X, Y) a <u>biconformable colouring</u> if $|V_{<\Delta}(X \backslash X_i)| \geq n_i - m_i$, $|V_{<\Delta}(Y \backslash Y_i)| \geq m_i - n_i$, and $\text{def}(G) \geq \sum_{i=1}^{s} |m_i - n_i|$, where X_i (resp. Y_i) is the set of vertices of X (resp. Y) coloured c_i. We say that G is <u>biconformable</u> if it has a biconformable vertex-colouring.

Suppose G is a regular, equibipartite graph having bipartition (X, Y). Prove that G is biconformable if and only if G has a vertex-colouring π using colours $c_1, ..., c_s$, where $s = \Delta(G) + 1$, such that $m_j = n_j$, where m_j (resp. n_j) is the number of vertices in X (resp. Y) coloured c_j. (Chetwynd and Hilton [88])

10. Prove that every biconformable balanced bipartite graph is conformable. (Chetwynd and Hilton [88])

11. Prove that C_8 is conformable but not biconformable.

12. Show that the graph given in Fig.2.2 is biconformable. Is this a Type 1 graph?

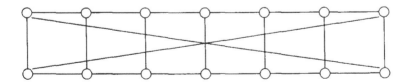

Figure 2.2

(Chetwynd and Hilton [88])

13.* Let G be a bipartite graph having $\Delta(G) \geq \frac{3}{14}(v(G) + 1)$. Then G is Type 2 if and only if G contains a balanced bipartite subgraph H, with $\Delta(H) = \Delta(G)$, which is not biconformable. (Remarks: The graph given in Fig. 2.2 shows that the bound $\frac{3}{14}$ cannot be lowered.) (Chetwynd and Hilton [88])

14. Let G be a bipartite graph having bipartition (X, Y). Let $\Delta_X = \max\{d(x)|x \in X\}$ and $\Delta_Y = \max\{d(y)|y \in Y\}$. Suppose $\Delta_X \geq \Delta_Y$ and that G has an edge xy such that $d(x) = \Delta_X$ and $d(y) = \Delta_Y$. Prove that $\chi_T(H) = \Delta(H) + 1$ where H is the line graph of G. (Sánchez-Arroyo [91])

CHAPTER 3

COMPLETE r-PARTITE GRAPHS

In this chapter we first determine the exact value of $\chi_T(G)$ for $G = K_n$ and $G = K_{m,n}$ (due to Behzad, Chartrand, and Cooper [67]). We next apply Lemma 2.6 to show that every complete r-partite graph satisfies the TCC. (This result is due to Yap [89]. It generalizes an earlier result of Rosenfeld [71].) Finally we apply Theorem 2.7 to give a complete classification of balanced complete r-partite graphs according to their total chromatic numbers (due to Bermond [74]).

§1. Complete graphs and complete bipartite graphs

We shall use the following edge-colouring of K_n very often. It will be referred as the <u>natural edge-colouring</u> of K_n. Let $V(K_n) = \{v_1, v_2, ..., v_n\}$.

Suppose n is odd. The vertices of a regular n-gon are taken as the vertices of K_n. The edges $v_2v_n, v_3v_{n-1}, v_4v_{n-2}, ...$ are parallel. The edges $v_3v_1, v_4v_n, v_5v_{n-1}, ...$ are also parallel and so on. We colour the first set of parallel edges by colour 1 and the second set by colour 2 and so on. This gives an n-edge-colouring of K_n.

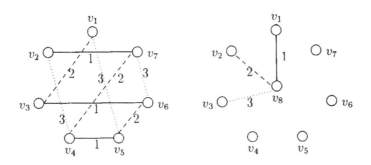

Figure 3.1

Suppose n is even. We first colour the edges of $K_n - v_n$ as above. We observe tha colour i is absent at vertex v_i, $i = 1, 2, ..., n - 1$. We then colour the edge $v_n v_i$ with colour i, $i = 1, 2, ..., n - 1$. This gives an $(n - 1)$-edge-colouring of K_n. For instance, we show in Figure 3.1 the cases $n = 7$ and $n = 8$.

The natural colouring of K_n has been known a few hundred years ago. It is called the GK_n colouring in Mendelsohn and Roasa [85].

Theorem 3.1

$$\chi_T(K_n) = \begin{cases} n & \text{if } n \text{ is odd,} \\ n + 1 & \text{if } n \text{ is even.} \end{cases}$$

Proof. Clearly $\chi_T(K_n) \geq \Delta(K_n) + 1 = n$. Let $V(K_n) = \{v_1, v_2, ..., v_n\}$.

Suppose n is odd. Then by Lemma 2.9, for each $p = 1, 2, ..., n$,

$$T_p = \left\{ v_{p+q} v_{p-q} \mid q = 1, 2, ..., \frac{n-1}{2} \right\} \cup \{v_p\},$$

where each of the numbers $p - q$ and $p + q$ is calculated modulo n, is a maximum independent set of elements in K_n. We further observe that $T_1, T_2, ..., T_n$ is a partition of $VE(K_n)$. Thus $\chi_T(K_n) \leq n$, and consequently $\chi_T(K_n) = n$.

Suppose n is even. Then from the facts that $\alpha_T(K_n) \leq \frac{n}{2}$ and $|VE(K_n)| = n + \frac{n(n-1)}{2} = \frac{n(n+1)}{2}$ it follows that $\chi_T(K_n) \geq n + 1$. Now by Lemma 2.1 and the result that $\chi_T(K_{n+1}) = n + 1$, we have $n + 1 \leq \chi_T(K_n) \leq \chi_T(K_{n+1}) = n + 1$. Consequently, $\chi_T(K_n) = n + 1$. //

Theorem 3.2

$$\chi_T(K_{m,n}) = \max(m, n) + 1 + \delta_{mn},$$

where δ_{mn} is the Kronecker delta function.

Proof. We assume that $m \geq n$ and we let $G = K_{m,n}$. Let the bipartition of G be (X, Y), where $X = \{x_1, x_2, ..., x_m\}$ and $Y = \{y_1, y_2, ..., y_n\}$. We define

$$A_p = \{y_i x_{i+p} \mid i = 1, 2, ..., n\} \text{ for } p = 1, 2, ..., m,$$

where the suffixes are taken modulo m. Then $A_1, A_2, ..., A_m$ are independent and $E(G) = A_1 \cup A_2 \cup ... \cup A_m$.

Suppose $m > n$. Then $\Delta(G) = m$ and $\chi_T(G) \geq m + 1$. For $p = 1, 2, ..., m$, we let $B_p = A_p \cup \{x_p\}$ and $B_{m+1} = Y$. The sets $B_1, B_2, ..., B_{m+1}$ are all independent sets and they form a partition of $VE(G)$. Hence $\chi_T(G) = m + 1$.

Suppose $m = n$. Then by Lemma 2.9, $\alpha_T(G) \leq m$. Thus from the facts that $|VE(G)| = m^2 + 2m = m(m + 2)$ and $\alpha_T(G) \leq m$ it follows that $\chi_T(G) \geq m + 2$. However, the sets $A_1, A_2, ..., A_m$ constructed above together with $A_{m+1} = X$ and $A_{m+2} = Y$ give a partition of $VE(G)$ into $m + 2$ independent sets. Hence $\chi_T(G) = m + 2$. //

The n-edge-colouring of $K_{n,n}$ n odd, where $A_1, ..., A_n$ are the n colour classes, given in the proof of Theorem 3.2 will be referred as the natural n-edge-colouring of $K_{n,n}$. We shall require the following result in Chapter 4 and Chapter 6.

Lemma 3.3 For any integer $n \geq 3$, there exists an n-edge-colouring of $K_{n,n}$, such that $K_{n,n}$ has a perfect matching receiving n distinct colours.

Proof. Our proof is constructive. Let (X, Y) be the bipartition of $K_{n,n}$, where $X = \{x_1, x_2, ..., x_n\}$ and $Y = \{y_1, y_2, ..., y_n\}$. We consider two cases separately.

Case 1. n is odd.

We observe that in the natural n-edge-colouring of $K_{n,n}$, the following is a perfect matching of $K_{n,n}$ coloured with colours $1, 2, ..., n$ respectively (here $n = 2m + 1$).

$$F = \{y_1 x_2, y_2 x_4, ..., y_m x_{2m}, y_{m+1} x_1, y_{m+2} x_3, ..., y_n x_n\}$$

Case 2. $n \geq 4$ is even.

From the construction of $A_1, ..., A_n$ given in the proof of Theorem 3.2, we notice that the following set of $n - 1$ (odd) independent edges receive respectively colours $2, 3, ..., n-1$ (here $n-1 = 2m+1$): $E_n' = \{y_1 x_3, y_2 x_5, ..., y_p x_{2p+1}, ..., y_m x_{n-1}, y_{m+1} x_2,$ $..., y_{n-2} x_{n-2}, y_{n-1} x_1\}$, where $2p + 1$ is calculated modulo $n - 1$. Now recolour each edge in E_n' by colour n, setting $\varphi(y_n x_3) = 2 = \varphi(x_n y_1)$, $\varphi(y_n x_5) = 3 = \varphi(x_n y_2), ..., \varphi(y_n x_{n-2}) = n-1 = \varphi(x_n y_{n-2}), \varphi(y_n x_1) = 1 = \varphi(x_n y_{n-1})$ and $\varphi(x_n y_n) = n$, we obtain an n-edge-colouring φ of $K_{n,n}$. We notice that the set of $n - 1$ independent edges constructed in Case 1 together with $y_n x_n$ form a perfect matching F of $K_{n,n}$ coloured with colours $1, 2, ... n$ respectively. //

(Remark: A parallel formulation of Lemma 3.3 already known in latin squares: "For any $n \geq 3$, one can construct a latin square of side n having a transversal." We give a proof of Lemma 3.3 here because the proof technique will be used later.)

§2. Complete r-partite graphs of odd order

Rosenfeld [71] proved that the TCC holds for balanced complete r-partite graphs. This result has recently been extended to all complete r-partite graphs by Yap [89].

Theorem 3.4 (Rosenfeld [71]; Yap [89]) For any complete r-partite graph G,

$$\chi_T(G) \leq \Delta(G) + 2.$$

Proof. Let $G = O_{p_1} + O_{p_2} + ... + O_{p_r}$, $p_1 \leq p_2 \leq ... \leq p_r$ and let $|V(G)| = n$. We have $\Delta(G) = n - p_1$. We take $S = V(O_{p_r})$. Then $|S| = p_r \geq n - \Delta(G) - 1$ and the result follows from Lemma 2.6.   //

We shall next determine the exact value $\chi_T(G)$ for complete r-partite graphs of odd order. The following two lemmas will be applied to prove Theorem 3.7.

Lemma 3.5 For each integer $t \geq 2$, every balanced complete t-partite graph G of even order has a perfect matching.

Proof. Let $V_1, V_2, ..., V_t$ be the partite sets of G. If t is even, then the union of maximum matchings in $G[V_1 \cup V_2], ..., G[V_{t-1} \cup V_t]$ is a perfect matching of G.

If t is odd, then $|V_i| = p$ is even. Let $V_i = X_i \cup Y_i$, $|X_i| = |Y_i| = \frac{p}{2}$, $i = 1, 2, 3$. Then the union of the maximum matchings in $G[X_1 \cup X_2]$, $G[Y_2 \cup Y_3], G[X_3 \cup Y_1]$, $G[V_4 \cup V_5], ..., G[V_{t-1} \cup V_t]$ is a perfect matching in G.   //

Remarks. Lemma 3.4 also follows immediately from Tutte's 1-factor theorem.

Lemma 3.6 For any positive integers n and p with $n > p$, we can always construct a circulant graph H of order n having degree $n - p - 1$ for any even n and for any odd n provided p is even.

Proof. The vertex set of H is the group $Z_n = \{0, 1, ..., n - 1\}$ under addition modulo n and the edge set of H is $\{(i, j) | i - j \in S\}$ where $S \subseteq Z_n$ is such that $0 \notin S$, $s \in S \Rightarrow -s \in S$ and $|S| = n - p - 1$. //

In applying Lemma 3.6, sometimes the vertices of H are labelled as $v_0, v_1, ..., v_{n-1}$ instead of $0, 1, ..., n - 1$.

The following theorem was proved independently by Chew and Yap [92], as well as Hoffman and Rodger [92]. Chew and Yap give a direct proof of this theorem and then use a similar proof technique to prove its edge-analogue, whereas Hoffman and Rodger first prove a theorem on the chromatic index of complete r-partite graphs of even order and then use it to deduce this theorem.

Theorem 3.7 (Chew and Yap [92]; Hoffman and Rodger [92]) Every complete r-partite graph G of odd order is Type 1.

Proof. Let $G = O_{p_1} + O_{p_2} + ... + O_{p_r}$, $p = p_1 = p_2 = ... = p_t < p_{t+1} \leq ... \leq p_r (1 \leq t \leq r)$. Let $\Delta = \Delta(G)$, $V_i = V(O_{p_i})$ and $V_i = \{v_{i0}, v_{i1}, ..., v_{i,p_i-1}\}$. By Theorem 3.2, we can assume that $r \geq 3$.

We shall prove that $G - V_1$ contains a matching M and that there exists $\cup E_i \subseteq E(\bar{G})$ such that $G^* = G - M + \cup E_i$ satisfies the following condition:

(1) $d_{G^*}(v) = \Delta$ for every $v \in V_1$ and $d_{G^*}(v) = \Delta - 1$ for every $v \in V(G^*) \setminus V_1$.

By Lemma 2.5, G^* has an edge-colouring π using Δ colours $1, 2, ..., \Delta$. Since $|G|$ is odd, each colour is missing at some vertices (at least one vertex). Now since each of the Δ minor vertices of G^* is of degree $\Delta - 1$, each colour j is missing at exactly one of these minor vertices. Thus we can extend π to a $(\Delta + 1)$-total-colouring of G by assigning colour $\Delta + 1$ to each element in $V_1 \cup M$ and putting $\pi(v) = j$ if colour j is missing at vertex v.

We next show that such a graph G^* exists. We consider two cases separately.

Case 1. p is even.

If $t \geq 3$, then by Lemma 3.5, we take M to be a perfect matching in $G[V_2 \cup ... \cup V_t]$ and if $t = 1$ we take $M = \phi$. Now for each $i > t$, we turn O_{p_i} into a circulant graph of degree $p_i - p - 1$. Suppose $t = 2$. Since $|G|$ is odd, there is an $m > t$ such that

$|V_m|$ is odd. For each $i \geq 3$, $i \neq m$, we turn O_{p_i} into a circulant graph of degree $p_i - p - 1$ (by Lemma 3.6). We can also turn O_{p_m} into a circulant graph H_m of degree $p_m - p - 1$ using the set of symbols $S = \{\pm 2, ..., \pm s\}$, where $s = (p_m - p + 1)/2$. We then add $p/2$ independent edges $v_{m0}v_{m1}, v_{m2}v_{m3}, ..., v_{m,p-2}v_{m,p-1}$ to H_m and delete the set of edges $M = \{v_{20}v_{m0}, v_{21}v_{m1}, ..., v_{2,p-1}v_{m,p-1}\}$ from G to obtain a required G^*.

Case 2. p is odd.

Let $W_1, W_2, ..., W_q$ be the odd partite sets of G other than V_1 and let $|W_1| \leq |W_2| \leq ... \leq |W_q|$. Since $|G|$ is odd, q must be even.

We take M to be the union of one maximum matching in each $G[W_{2i-1} \cup W_{2i}]$, $i = 1, 2, ..., q/2$. Using the proof technique of Case 1, we can add a set of edges E_j to $G[W_j]$ such that for each odd j, $G[W_j] + E_j$ is a circulant graph of degree $|W_j| - p$ ($E_j = \phi$ if $|W_j| = p$) and for each even j, if $|W_{j-1}| < |W_j|$, then for every M-saturated vertex $w \in W_j$, w is of degree $|W_j| - p$ in $G[W_j] + E_j$ and for every M-unsaturated vertex $w \in W_j$, w is of degree $|W_j| - p - 1$ in $G[W_j] + E_j$. (We first turn $G[W_j]$ into a circulant graph H'_j of degree $|W_j| - p$ using the set of symbols $S = \{\pm 1, ..., \pm s\}$, where $s = (|W_j| - p)/2$ and then delete the edges $v'_{j0}v'_{j1}, ..., v'_{j,k-2}v'_{j,k-1}$, where $W_j = \{v'_{j0}, ..., v'_{j,|W_j|-1}\}$ and $k = |W_j| - |W_{j-1}|$. Here $v'_{j0}, ..., v'_{j,k-1}$ are the M-unsaturated vertices and $v'_{j,k}, ..., v'_{j,|W_j|-1}$ are the M-saturated vertices.) For each of the even partite sets V_i, we can turn O_{p_i} into a circulant graph of degree $p_i - p - 1$. The resultant graph G^* satisfies condition (1).   //

§3. Bermond's theorem

We shall require the following two lemmas to prove Bermond's theorem which determines the exact total chromatic number of all balanced complete r-partite graphs.

Lemma 3.8 For any integer $r \geq 3$,

$$\chi_T(K(r,2)) = 2r - 1.$$

Proof. Suppose r is odd. Let φ be an r-total-colouring of K_r with colours $1, 2, ..., r$. We first observe that

$$K(r,2) \simeq (K_r + K_r) - F,$$

where $K_r + K_r$ denotes the join of K_r with K_r and F is a 1-factor (perfect matching) of $K_r + K_r$ such that each edge in F joins two vertices receiving the same colour in the two copies of K_r. Since the set of uncoloured edges in $(K_r + K_r) - F$ induces a bipartite graph having maximum degree $r - 1$, these edges can be coloured with colours $r + 1, ..., 2r - 1$. Hence $\chi_T(K(r,2)) = 2r - 1$.

Suppose r is even. We colour the edges of the two copies of K_r with $r - 1$ colours $1, 2, ..., r - 1$. Now the set of edges joining the two copies of K_r form a complete bipartite graph $K_{r,r}$. By Lemma 3.3, the edges of $K_{r,r}$ can be coloured with r colours $r, r + 1, ..., 2r - 1$ such that it has a perfect matching $F = \{x_1 y_1, ..., x_n y_n\}$ say, receiving distinct colours $r, r + 1, ..., 2r - 1$. We can now use colour $r - 1 + i$ to colour the two vertices x_i and y_i. Thus $K(r,2) \simeq (K_r + K_r) - F$ has a total-colouring using $2r - 1$ colours and consequently $\chi_T(K(r,2)) = 2r - 1$. $\quad //$

Lemma 3.9 Let r be an odd integer. Let G be the graph having

$$V(G) = \{x_1, x_2, ..., x_r\} \cup \{y_1, y_2, ..., y_r\},$$

and $\quad E(G) = \{x_1 x_2, ..., x_r x_1\} \cup \{y_1 y_2, ..., y_r y_1\} \cup \{x_1 y_r, x_1 y_2, ..., x_r y_{r-1}, x_r y_1\}$.

Then G has two edge-disjoint Hamilton cycles.

The following is an example of a graph given in Lemma 3.9 in which $r = 7$.

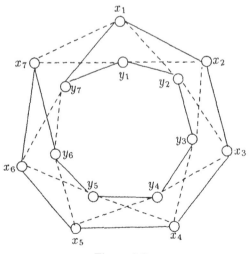

Figure 3.2

Proof of Lemma 3.9. We note that $x_1x_2...x_rx_1$ and $y_1y_2...y_ry_1$ form two cycles in G and that G is regular of degree 4.

We establish two edge-disjoint Hamilton cycles explicitly as follws:

$$C_1 : x_1x_2...x_ry_{r-1}y_{r-2}...y_2y_1y_rx_1, \quad C_2 = G - E(C_1). \quad //$$

(Note that Lemma 3.9 is also true if $r \geq 4$ is even.)

Theorem 3.10 (Bermond [74])

$$\chi_T(K(r,n)) = \begin{cases} \Delta(K(r,n)) + 2 & \text{if } r = 2 \text{ or } r \text{ even and } n \text{ odd} \\ \Delta(K(r,n)) + 1 & \text{otherwise.} \end{cases}$$

Proof. The case $r = 2$ has been settled in Theorem 3.2. Hence we assume that $r \geq 3$.

We shall now first prove that if $r \geq 4$ is even and n is odd, then $G = K(r,n)$ is Type 2. Note that $\Delta(G) = (r-1)n$. Now $e(\bar{G}) + \alpha'(\bar{G}) = r\frac{n(n-1)}{2} + r\frac{n-1}{2} = \frac{r}{2}(n^2 - 1) \not\geq \frac{rn}{2}[rn - (r-1)n]$. Hence, by Theorem 2.7, G is Type 2.

It remains to prove that if r is odd or if both $r \geq 4$ and n are even, then G is Type 1. By Theorem 3.7 (in fact we apply only Case 2 in the proof of Theorem 3.7) we need only to consider the case $r \geq 3$ and n is even.

Let $V_1, V_2, ..., V_r$ be the partite sets and let $V_i = \{v_{i1}, v_{i2}, ..., v_{in}\}, i = 1, 2, ..., r$. Let $m = \frac{n}{2}$. Then

$$G_1 = G[v_{11}, v_{12}, v_{21}, v_{22}, ..., v_{r1}, v_{r2}] \simeq ... \simeq G_m$$

$$= G[v_{1,n-1}, v_{1,n}, v_{2,n-1}, v_{2,n}, ..., v_{r,n-1}, v_{r,n}] \simeq K(r,2)$$

and thus, by Lemma 3.8, they can be simultaneously total coloured with the same set of colours $\{1, 2, ..., 2r - 1\}$ such that vertices in the same partite set V_j receive the same colour. Let E' be the set of edges in $G_1 \cup ... \cup G_m$ which have been coloured and let $G' = G - E'$. Then G' is regular of degree $(r-1)n - 2(r-1) = (r-1)(n-2)$. We now consider two cases separately:

Case 1. $r \geq 4$ is even.

Let K_r^* be the complete graph having vertex set $\{V_1, V_2, ..., V_r\}$ (each V_i is considered as a vertex). Since $\chi'(K_r^*) = r - 1$ and for any $i \neq j$, $G'[V_i \cup V_j]$ is regular of degree $n - 2$, for each edge $V_i V_j$ in a colour class of an $(r - 1)$-edge-colouring of K_r^* we replace it by $n - 2$ perfect matchings in $G'[V_i \cup V_j]$. Hence G' can be edge-coloured with $(r - 1)(n - 2)$ colours. Consequently $\chi_T(G) \leq (r - 1)(n - 2) + 2r - 1 = (r - 1)n + 1 = \Delta + 1$ and thus $\chi_T(G) = \Delta + 1$.

Case 2. $r \geq 3$ is odd.

Let $X_i = \{v_{i1}, ..., v_{im}\}, Y_i = \{v_{i,m+1}, ..., v_{in}\}, i = 1, 2, ..., r$. By Lemma 3.9, we can find the first $2(n - 2)$ perfect matchings of G' in the cyclic order $(12...r)$ (the first four perfect matchings are the edges in the two edge-disjoint Hamilton cycles $v_{11} v_{21} ... v_{r1} v_{r-1,m+1} v_{r-2,m+1}, ..., v_{2,m+1} v_{1,m+1} v_{11}$ and $v_{11} v_{2,m+1} v_{31} v_{4,m+1} ... v_{r,m+1}$ $v_{r-1,1} v_{r-2,m+1} ... v_{31} v_{2,m+1} v_{11}$ of $G[\{v_{11}, v_{1,m+1}, v_{21}, v_{2,m+1}, ..., v_{r1}, v_{r,m+1}\}]$, see the two edge-disjoint Hamilton cycles of Fig 3.2.), the second $2(n - 2)$ perfect matchings of G' in the cyclic order $(135...)$, the third $2(n - 2)$ perfect matchins of G' in the cyclic order $(147...)$ (if $3 \nmid r$) or in the cyclic order $(147...)(258...)(369...)$ (if $3|r$) and so on. Hence G' can be edge-coloured with $\frac{(r-1)}{2} 2(n - 2) = (r - 1)(n - 2)$ colours. Consequently $\chi_T(G) \leq (r - 1)(n - 2) + 2r - 1 = \Delta + 1$ and thus $\chi_T(G) = \Delta + 1$. $//$

In Exercise 3, we shall see that there are many unbalanced complete r-partite graphs of even order which are Type 1. However, applying Theorem 2.7, we can prove that for any $k \geq 3$ if $\bar{G} = 2k K_3 \cup K_4$, then G is Type 2. (Let $n = 3k + 2$. We observe that $\Delta(G) = 2n - 3$, $e(\bar{G}) + \alpha'(\bar{G}) = (6k + 6) + (2k + 2) = 8k + 8 \geq 9k + 6 = 3n = n(2n - \Delta)$ only if $k \leq 2$.) Hence there are also many unbalanced complete r-partite graphs of even order which are Type 2.

To end this chapter, we pose the following conjecture.

Conjecture. Suppose $\bar{G} = K_{p_1} \cup K_{p_2} \cup ... \cup K_{p_r}$, where $2n = p_1 + p_2 + ... + p_r$ and $p_1 \leq p_2 \leq ... \leq p_r$. Then the unbalanced complete r-partite graph G is Type 1 if and only if $e(\bar{G}) + \alpha'(\bar{G}) \geq p_1 n$.

Remarks. Further results on the total chromatic number of unbalanced complete r-partite graphs of even order have been obtained by Hoffman and Rodger [-a] as well as by Shen Minggang [-a] (Shen [-a] claimed that he had proved the above conjecture. However, there are several serious errors in his paper.)

Exercise 3

1. Let G be a multigraph whose underlying graph is a complete graph. Prove that

$$\chi_T(G) \leq \Delta(G) + \mu(G) + 1.$$

2.⁻ Give an example of a multigraph G such that

$$\chi_T(G) < \Delta(G) + \mu(G).$$

3.⁻ List all the elements in the sets $B_1, B_2, \ldots$ and B_6 for the case $m = 4$ and $n = 6$ constructed in the proof of Theorem 3.2.

4. Suppose G is a graph on at least two vertices and $\chi(G) + \chi'(G) = \chi_T(G)$. Prove that G is bipartite. (Behzad, Chartrand and Cooper [67])

5. Suppose G is a graph of order p. Prove that
   (i) $p + 1 \leq \chi_T(G) + \chi_T(\bar{G}) \leq 2p$; and
   (ii) $p + 1 \leq \chi_T(G)\chi_T(\bar{G}) \leq p^2$.
   (Cook [74]; Wang and Zhang [87])

6. Let $G = O_{p_1} + O_{p_2} + \ldots + O_{p_r}$, where $p_1 < p_2 \leq \ldots \leq p_r$. Prove that G is Type 1.

7. Let $G = O_{p_1} + O_{p_2} + O_{p_3}$, where $p_1 = p_2 < p_3$. Prove that G is Type 1.

8.⁺ Prove that $G = O_5 + O_5 + O_6 + O_6$ is Type 1.

9.⁺ Let $G = O_{p_1} + O_{p_2} + \ldots + O_{p_r}$ be of even order with $p = p_1 = p_2 < p_3 \leq \ldots \leq p_r$, where $p_{r-1} \geq p + 2$ and $p_3 + p_4 + \ldots + p_{r-1} \geq 2p + p_r - 3$. Prove that G is Type 1. Hence deduce that

$$G = O_p + O_p + O_{p+1} + O_{p+1} + O_{p+2} + O_{p+2}$$

   and $$G' = O_p + O_p + O_{p+2} + O_{p+2} + O_{p+2} + O_{p+2}$$

   are Type 1.
   (Hint: You may apply the fact that if G^* is a graph of odd order $2n + 1$ having r major vertices, $\delta(G^*) \geq n + r - 2$, and G^* has s minor vertices $z_1, \ldots, z_s$ such that $d(z_i) \leq \Delta(G^*) - 2$ for each $i = 1, 2, \ldots, s$ and $\sum_{i=1}^{s} (\Delta(G^*) - d(z_i)) \geq 2r$ then G^* is Class 1 - see Chew and Yap [-b].)

10. Applying Lemma 2.6, give an alternative proof of Theorem 3.2.

CHAPTER 4

GRAPHS OF LOW DEGREE

In this chapter we shall prove that the TCC holds for graphs having maximum degree at most 4.

§1. Graphs G having $\Delta(G) = 3$

It is clear that the TCC holds for any graph G having $\Delta(G) \leq 2$. In fact, from Exercise 2(1), we can give a complete classification of Type 1 and Type 2 graphs G having $\Delta(G) \leq 2$. As for graphs G having $\Delta(G) = 3$, the TCC was proved by Rosenfeld [71] and independently Vijayaditya [71]. Rosenfeld's proof is by induction on the order of G and, as a result, many cases have to be considered separately when G is 2-edge-connected. Vijayaditya's proof is by decomposition of $E(G)$ into an edge-disjoint union of a 2-factor (i.e. cycles) and a 1-factor (i.e. perfect matching). We shall present a short proof of Rosenfeld-Vijayaditya result in this section. Vijayaditya's proof will also be presented.

We shall apply the following two useful lemmas to prove several theorems in this chapter and in Chapter 7 and Chapter 8.

The first lemma embodied an idea originally due to Kostochka [77b]. This explicit statement first appeared in McDiarmid and Sànchez-Arroy [93]. It is further employed in Hind's paper [90] as well as Hilton and Hind's paper [93].

We first define a few terms. Suppose $G = (V, E)$ is a graph, $W \subseteq V$ and $E' \subseteq E$. A (proper) vertex-colouring ϕ of $G[W]$ is called a partial vertex-colouring of G. A (proper) edge-colouring ψ of $G[E']$ is called a partial edge-colouring of G. A colouring π of $W \cup E'$ is said to be a partial total-colouring of $W \cup E'$ if $\pi|_W$ is a partial vertex-colouring of G, $\pi|_{E'}$ is a partial edge-colouring of G and for each $v \in W$, $e \in E'$, if v and e are incident, then $\pi(v) \neq \pi(e)$.

Lemma 4.1 Suppose $G = (V, E)$ is a graph having $\Delta(G) \leq 2$. Let $W \subseteq V$ and let $\phi : W \mapsto \{1, 2, 3, 4\}$ be a partial vertex-colouring of G. Then there exists a partial edge-colouring $\psi : E \mapsto \{1, 2, 3, 4\}$ such that $\phi \cup \psi$ is a partial total-colouring of $W \cup E$ and for any edge $e = xy$ of G, if $\psi(e) = 4$ then $x, y \in W$.

Proof. It suffices to consider a cycle-component $C = v_1 v_2 ... v_n v_1$ of G.

We extend ϕ to a vertex-colouring of G by setting $\phi(v) = 4$ for any $v \in V \setminus W$. (Note that now ϕ need not be a proper vertex-colouring of G.) We consider two cases:

Case 1. Suppose $\phi(v_{i+2}) \neq \phi(v_i)$ for some i (taking modulo n if $i + 2 > n$), say $\phi(v_2) \neq \phi(v_n)$. Without loss of generality, we assume that $\phi(v_n) \neq 4$. We first set $\psi(v_1 v_2) = \alpha_1 = \phi(v_n)$. Then $\psi(v_1 v_2) \neq \phi(v_1)$ since $\phi : W \to \{1, 2, 3, 4\}$ is a partial vertex-colouring of G. We next set $\psi(v_2 v_3) = \alpha_2 \in \{1, 2, 3, 4\}$ such that $\alpha_2 \neq \phi(v_2)$, $\phi(v_3)$, α_1. Clearly if $\alpha_2 = 4$, then $\phi(v_2) \neq 4$, $\phi(v_3) \neq 4$ and thus $v_2, v_3 \in W$. We continue colouring the edges $v_3 v_4, v_4 v_5, ..., v_{n-1} v_n$ in this way. We notice that for $e = v_i v_{i+1}$ if $\psi(e) = 4$, then $\phi(v_i) \neq 4$ and $\phi(v_{i+1}) \neq 4$ and thus v_i, $v_{i+1} \in W$. We last set $\psi(v_n v_1) = \alpha_n \in \{1, 2, 3, 4\}$ such that $\alpha_n \neq \phi(v_n)$, $\phi(v_1)$, α_{n-1}. Again if $\alpha_n = 4$, then $\phi(v_1) \neq 4$ and thus $v_n, v_1 \in W$.

Case 2. Suppose $\phi(v_{i+2}) = \phi(v_i)$ for all $i = 1, 2, ..., n$. Then C is either vertex-monochromatic (and thus $\phi(v_i) = 4$ for all i) or vertex-coloured with two colours (and thus C is an even cycle). In the case that all the vertices of C were coloured with colour 4, we can colour the edges with colours $1, 2, 3$. In the case that C is an even cycle whose vertices were alternately coloured with colours α and β, we can colour the edges of C with colours γ, $\delta \in \{1, 2, 3, 4\}$, where γ, $\delta \neq \alpha, \beta$. Clearly the requirement that for any edge $e = xy$ such that $\psi(e) = 4$, then $x, y \in W$ is satisfied. //

Lemma 4.2 Let $G = (V, E)$ be a graph, $W \subseteq V$, and $\phi : W \mapsto \{1, 2, 3, 4\}$ a partial vertex-colouring of G. Suppose G has a set of q matchings E' such that $H = G - E'$ has $\Delta(H) \leq 2$. Then G has an edge-colouring $\psi : E \mapsto \{1, 2, ..., q + 4\}$ such that $\phi \cup \psi$ is a partial total-colouring of $W \cup E$ and for any edge $e = xy$ of G, if $\psi(e) = 4$ then $x, y \in W$.

Proof. Apply Lemma 4.1 to H. We complete the proof by putting back the q matchings E' and colour them with colours $5, ..., q + 4$. We note that for any edge $e' = x'y'$ in E', if both the two ends x' and y' are in W, then since ϕ is a partial vertex-colouring, $\phi(x') \neq \phi(y')$. Hence ψ is a partial total-colouring of $W \cup E$. //

The following lemma will also be required. A proof of this lemma can also be found in D. König [50]; or in Behzad, Chartrand and Lesniak-Foster [79;p.165].

Lemma 4.3 (Petersen, 1891) Suppose G is a cubic graph which contains no cut-edge. Then $E(G)$ can be decomposed into a union of edge-disjoint cycles and a 1-factor of G.

Proof. It suffices to show that G has a 1-factor. Suppose otherwise. Then by Tutte's 1-factor theorem, there exists a proper subset S of $V(G)$ such that the number of odd components of $G - S$ exceeds $|S|$. Let $G_1, G_2, ..., G_n$ be the odd components of $G - S$. Since G has no cut-edge, there are at least two edges joining G_i to S for each $i = 1, 2, ..., n$.

Suppose for some $i = 1, 2, ..., n$ there are exactly two edges joining G_i and S. Since $|G_i|$ is odd, G_i will then have an odd number of vertices of odd degree, which is false. Hence, for each $i = 1, 2, ..., n$, there are at least three edges joining G_i and S. Consequently the total number of edges joining $V(G_1) \cup ... \cup V(G_n)$ and S is at least $3n$. However, since each vertex of S has degree 3, the number of edges joining $V(G_1) \cup ... \cup V(G_n)$ and S is at most $3|S|$. Hence $3|S| \geq 3n$, a contradiction. //

In the last part of the proof of Lemma 4.5 we shall refer to the following specific colouring π of a cycle $C_n : x_1 \, x_2 \ldots x_n \, x_1$.

Case 1. Suppose n is even. Then

$$\pi(x_i) = 1 \quad \text{or} \quad 2 \quad \text{according as } i \text{ is odd or even,}$$
$$\pi(x_i x_{i+1}) = 3 \quad \text{or} \quad 4 \quad \text{according as } i \text{ is odd or even.}$$

Case 2. Suppose n is odd. Then

$$\pi(x_n) = 3 \quad \text{and} \quad \pi(x_i) = 1 \quad \text{or} \quad 2 \qquad \text{according as } i \neq n \text{ is odd or even,}$$
$$\pi(x_n x_1) = 2 \quad \text{and} \quad \pi(x_i x_{i+1}) = 3 \quad \text{or} \quad 4 \qquad \text{according as } i \neq n \text{ is odd or even.}$$

For odd n, the three vertices x_{n-1}, x_n, x_1 will be called <u>specified vertices</u>. It is not difficult to see that the total-colouring π of C_n given above remains a total-colouring of C_n if the colours assigned to x_i and $x_i x_{i+1}$ are interchanged for some of the non-specified vertices x_i.

Lemma 4.4 Let G be a connected graph having $\Delta(G) = 3$. If G is not regular or G contains a cut-edge, then $\chi_T(G) \leq 5$.

Proof. The proof is by induction on $|G|$. Suppose G contains a vertex x where $d(x) \leq 2$. Then it is not difficult to show that any 5-total-colouring π of $G - x$ can be extended to a total-colouring of G using the same set of five colours. (Details of proof are left as an exercise.)

Suppose G is regular and G contains a cut-edge $e = uv$. Then $G - e$ has two components G_1 and G_2, where G_1 contains u, G_2 contains v, and $d_{G_1}(u) = 2 = d_{G_2}(v)$. Let π_1 and π_2 be respectively two total-colourings of G_1 and G_2 using the five colours 1, 2, 3, 4 and 5. By permutation of colours in π_1, if necessary, we can assume that $\pi_1(u)$ is not a colour presenting at vertex v and the two colours assigned to the two edges incident with u in G_1 are the same as the two colours assigned to the two edges incident with v in G_2. It is not difficult to see that π_1 and π_2 can now be combined and extended to a total-colouring of G using the five colours $1, 2, 3, 4$ and 5.   //

Lemma 4.5 If G is a connected cubic graph which contains no cut-edge, then $\chi_T(G) \leq 5$.

First proof. Let φ be a proper vertex-colouring of $W = V(G)$ using four colours $1, 2, 3$ and 4. By Lemma 4.3, $E(G)$ can be decomposed into a union of edge-disjoint cycles and a 1-factor F. Let $G' = G - F$. Now by Lemma 4.1, φ can be extended to a (partial) total-colouring of G' using the same set of four colours $\{1, 2, 3, 4\}$. Finally, φ can be further extended to a total-colouring of G using five colours $\{1, 2, 3, 4, 5\}$ by putting $\varphi(e) = 5$ for all $e \in F$.   //

Second proof. By Lemma 4.3, $E(G)$ can be decomposed into a union of edge-disjoint cycles $C^1, C^2, ...$, and a 1-factor F.

We first show that we can label the vertices of $C^1, C^2, \ldots$ so that $C^i : x_1^i x_2^i \ldots x_{n_i}^i x_1^i$ and for any two odd cycles C^i and C^j the following condition is satisfied :

(1) $$x_1^i x_1^j, \; x_{n_i-1}^i x_{n_j-1}^j \notin E(G).$$

Step 1. If none of $C^1, C^2, \ldots$ is an odd cycle, we label the vertices of each cycle C^i as above starting with any vertex x_1^i in C^i. Otherwise we let $C^1, C^2, \ldots$ be the odd cycles and we also label one vertex of C^1 as x_1^1.

Step 2. If x_1^1 is adjacent to a vertex of an unlabelled odd cycle we go to Step 3. If x_1^1 is not adjacent to a vertex of an unlabelled odd cycle we label the other vertices of C^1 in any one of the two cyclic orders. Now if $x_{n_1-1}^1$ is adjacent to a vertex of an unlabelled odd cycle, we go to Step 4. Otherwise we go back to Step 1 to label the vertices of another unlabelled odd cycle C^2.

Step 3. When x_1^1 is adjacent to a vertex of an unlabelled odd cycle, we renumber this odd cycle as C^2, we let $x_{n_2-1}^2$ in C^2 be the vertex adjacent to x_1^1, and we label the other vertices of C^2 in any one of the two cyclic orders starting with $x_{n_2-1}^2$. We then go back to Step 2 to consider the vertex x_1^2 with the objective that condition (1) be satisfied.

Step 4. When x_1^1 is not adjacent to a vertex of an unlabelled odd cycle but $x_{n_1-1}^1$ is adjacent to a vertex, say y, of an unlabelled odd cycle, we renumber this odd cycle as C^2, and we label y as x_1^2. We label the other vertices of C^2 in any one of the two cyclic orders starting with x_1^2. We go back to Step 2 to consider $x_{n_2-1}^2$ with the objective that condition (1) be satisfied.

Step 5. Since G is finite, this algorithm will eventually come to an end. We then start with an unlabelled odd cycle and repeat the entire procedure from Step 2. After all the odd cycles have been labelled we can label the even cycles in any way.

Let π be the specific total-colouring of all the cycles $C^1, C^2, \ldots$ using colours 1, 2, 3 and 4 in the way as described earlier in this section. We shall finally modify π to a total-colouring of G using the five colours 1, 2, 3, 4 and 5 as follows : Let $x_h^k x_q^p \in F$. If $\pi(x_h^k) \neq \pi(x_q^p)$, we put $\pi(x_h^k x_q^p) = 5$. If $\pi(x_h^k) = \pi(x_q^p)$, we consider two cases separately :

Case i. Suppose one of the vertices x_h^k and x_q^p, say x_h^k, is not a specified vertex. We put $\pi(x_h^k x_q^p) = 5$ and we interchange the colours assigned to x_h^k with that of $x_h^k x_{h+1}^k$.

Case ii. Suppose both x_h^k and x_q^p are specified vertices. Then by the above labelling procedure, $\pi(x_h^k) = \pi(x_q^p) = 3$. We now put $\pi(x_h^k x_q^p) = 1$ and recolour x_h^k by colour 5.   //

Similar to vertex-colourings and edge-colourings of a graph, in the study of total-colourings of a graph G we need only to consider the case that G is connected. Hence from Lemma 4.4 and Lemma 4.5 we have

Theorem 4.6   For any graph G having $\Delta(G) = 3$,

$$\chi_T(G) \le 5.$$

§2. Graphs G having $\Delta(G) = 4$

Kostochka [77] proves that the TCC also holds for multigraphs G having $\Delta(G) = 4$. His result will be presented in this section.

We shall require the following lemmas to prove Kostochka's result. A proof of Lemma 4.7 can also be found in Behzad, Chartrand and Lesniak-Foster [79;p.166].

Lemma 4.7 (Petersen, 1891) If G is a $2n$-regular multigraph for some $n \ge 1$, then G is 2-factorizable.

Proof. We may assume that G is connected. Hence, G contains an eulerian circuit C.

Let $V(G) = \{v_1, v_2, ..., v_p\}$. We define a bipartite graph H with partite sets $U = \{u_1, u_2, ..., u_p\}$ and $W = \{w_1, w_2, ..., w_p\}$, where

$$E(H) = \{u_i w_j | v_j \text{ immediately follows } v_i \text{ on } C\}.$$

The graph H is n-regular and so by König's theorem (see Yap [86;p.11]), can be edge-decomposed into n factors $F_1, F_2, ..., F_n$.

Corresponding to each F_k is a permutation π_k on $\{1, 2, ..., p\}$, defined by $\pi_k(i) = j$ if $u_i w_j \in F_k$. Let π_k be expressed as a product of disjoint permutation cycles.

Then from the construction of $E(H)$, it follows that each permutation cycle of π_k has length at least 3.

Finally, each permutation cycle in π_k gives rise to a cycle in G, and the product of all disjoint permutation cycles in π_k produces a 2-factor in G. Since the 1-factors $F_1, F_2, ..., F_n$ in H are mutually edge-disjoint, the resulting 2-factors form a 2-factorization of G. //

Lemma 4.8 Let φ be a (proper) vertex-colouring of a cycle C_n. Let $\mathcal{C}$ be a set of colours such that $|\mathcal{C}| = 4$. Then φ can be extended to a total-colouring of C_n such that $\varphi(e) \in \mathcal{C}$ for any edge e of C_n.

Proof. Let $C_n : v_1 v_2 \ldots v_n v_1$. If the vertices of C_n were coloured with two colours α, β say, then C_n is an even cycle. Hence the edges of C_n can be coloured with two colours $\gamma, \delta (\neq \alpha, \beta)$ from $\mathcal{C}$.

Suppose φ uses $k \geq 3$ colours to colour the vertices of C_n. Without loss of generality, we assume that $\varphi(v_1) \neq \varphi(v_3)$. We then colour the edges of C_n as follows:

$$\varphi(v_2 v_3) = \begin{cases} \varphi(v_1) & \text{if } \varphi(v_1) \in \mathcal{C} \\ a \in \mathcal{C} \setminus \{\varphi(v_2), \varphi(v_3)\} & \text{if } \varphi(v_1) \notin \mathcal{C}, \end{cases}$$

$$\varphi(v_i v_{i+1}) = b \in \mathcal{C} \setminus \{\varphi(v_i), \varphi(v_{i+1}), \varphi(v_{i-1} v_i)\} \quad \text{for } 3 \leq i \leq n,$$

and $\varphi(v_1 v_2) = c \in \mathcal{C} \setminus \{\varphi(v_1), \varphi(v_2), \varphi(v_n v_1), \varphi(v_2 v_3)\}$. (Note that $\mathcal{C} \neq \{\varphi(v_1), \varphi(v_2),$ $\varphi(v_n v_1), \varphi(v_2 v_3)\}$ by the choice of $\varphi(v_2 v_3)$.) //

Theorem 4.9 (Kostochka [77]) For any multigraph G having $\Delta(G) = 4$, $\chi_T(G) \leq 6$.

Proof. By Lemma 2.1 and Theorem 2.2 we can assume that G is 4-regular. Since the TCC holds for K_5, we may assume that G is not a complete graph. Hence by Brooks' theorem, G has a vertex-colouring φ using four colours, say 1, 2, 3 and 4.

By Lemma 4.7, G is an edge-disjoint union of two 2-factors F_1 and F_2. By Lemma 4.8, φ can be extended to a total-colouring of F_1 using the same four colours 1, 2, 3 and 4. Now for each $v \in V(G)$ we define $A(v) = \{\varphi(e) | e = uv \in E(F_1)\}$.

A vertex u that lies on an odd cycle of F_2 will be called a _t-vertex_ if there exists $v \in V(G)$ such that $uv \in E(F_2)$ and $|\{\varphi(v)\} \cup A(u) \cup A(v)| \le 3$.

It is not difficult to verify that

(2) if u and v lie on an odd cycle of F_2 and $uv \in E(F_2)$ is such that $|A(u) \cup A(v)| \le 3$, then at least one of u and v is a t-vertex.

If C is an odd cycle in F_2, then clearly C contains at least two adjacent vertices u and v such that $|A(u) \cup A(v)| \le 3$. (Let $C : v_1 v_2 ... v_n$. Suppose otherwise. Then by permutation of colours, we can assume that $A(v_1) = \{1,2\}$. Hence $A(v_2) = \{3,4\}$, ..., $A(v_{n-1}) = \{3,4\}$. Now either $|A(v_1) \cup A(v_n)| \le 3$ or $|A(v_{n-1}) \cup A(v_n)| \le 3$, a contradiction.) Hence, by (2), C contains at least one t-vertex. In fact, we shall prove that

(3) every odd cycle $C : x_1 x_2 \ldots x_n x_1$ in F_2 contains at least two t-vertices.

Suppose to the contrary that C contains only one t-vertex x_i say. Without loss of generality, we can assume that $A(x_i) = \{1,2\}, A(x_{i+1}) = \{1,3\}$ and $\varphi(x_i) = 4$. Since x_i is the only t-vertex in C, by (2) we must have $A(x_{j+1}) = \{1,2,3,4\} \setminus A(x_j)$ for every $j = i+1, i+2, \ldots, n, 1, \ldots, i-3, i-2$. Hence $A(x_{i-1}) = \{2,4\}$ and $|A(x_i) \cup A(x_{i-1}) \cup \{\varphi(x_i)\}| = |\{1,2,4\}| = 3$ from which it follows that x_{i-1} is also a t-vertex, a contradiction.

We next produce an algorithm to select a maximal subset T of the set of t-vertices satisfying

(4) each odd cycle of F_2 contains exactly one vertex in T, and $G_1[T]$ contains no odd cycles, where $G_1 = G[E(F_1)]$.

Step 1. If F_2 contains no odd cycle, we put $T = \phi$. Otherwise we let C^1 be an odd cycle in F_2 and we select any t-vertex z_1 in C^1 and we call z_1 a _t_0-vertex_. Let $C^{(1)}$ be the cycle in F_1 containing z_1. We give a cycle orientation of $C^{(1)}$ so that $C^{(1)} : z_1 z_2 \ldots z_n z_1$.

Step 2. Suppose each of the vertices $z_1, z_2, \ldots, z_j, j < m$ has been determined as to whether it is a t_0-vertex. If z_{j+1} is a t-vertex and it lies on an odd cycle C of F_2 such that $V(C) \cap T = \phi$ we also call z_{j+1} a t_0-vertex. Otherwise z_{j+1} is called a _usual-vertex_.

Step 3. If at least one of the vertices $z_1, z_2, \ldots, z_m$ is a usual-vertex, or if m is even, then we put all the t_0-vertices in $\{z_1, z_2, \ldots, z_m\}$ into T. Otherwise we put all the vertices of $C^{(1)}$ into T, except z_2. In this later case we also call z_2 a usual-vertex. We now go to Step 1 to consider any t-vertex w_1 of an odd cycle C^2 in F_2 which contains no vertices in T.

Finally, we denote the odd cycles of F_2 by $C^1, C^2, \ldots$ and label the vertices of C^i such that $C^i : x_1^i x_2^i \ldots x_{2n_i+1}^i x_1^i$ where $x_1^i \in T$. By (4) above, we can recolour the vertices in T using colours 5 and 6. The edges of the odd cycles $C^1, C^2, \ldots$ are coloured in the following way. We let $\{\alpha, \beta\} = \{5, 6\}$. If $\varphi(x_1^i) = \alpha$, we put $\varphi(x_{2s}^i x_{2s+1}^i) = \alpha$ for all $1 \leq s \leq n_i$ and we put $\varphi(x_1^i x_2^i) = \gamma$ where $\{\gamma\} = \{1, 2, 3, 4\} \setminus (A(x_1^i) \cup A(x_2^i) \cup \{\varphi(x_2^i)\})$. The other edges of C^i are coloured with colour β. The edges of the even cycles are coloured with colours 5 and 6. We have thus obtained a total-colouring φ of G using the six colours 1, 2, 3, 4, 5, 6. $//$

Remarks.

(i) We can give a slightly shorter proof of Theorem 4.9 which is similar to the first proof of Lemma 4.5 (see Exercise 4(8)).

(ii) Hamilton and Hilton [91] give a list of all graphs of maximum degree 3 and order at most 16 which are critical with respect to the total chromatic number in the sense that $\chi_T(G) = 5$ and $\chi_T(G - e) = 4$ for any edge e of G.

(iii) Kostochka also proves, in his Ph.D. thesis (1978) that the TCC is true for multigraphs G having $\Delta(G) = 5$. His original proof was long (about 20 pages). Recently he has given a slightly shorter proof of this result (10 pages); see Kostochka [-a]).

Exercise 4

1.⁻ Supply details to paragraph one in the proof of Lemma 4.4.

2.⁻ Let G be a multigraph having $\Delta(G) = 3$. Prove that $\chi_T(G) \leq 5$.

3. Determine the total chromatic number of the Petersen graph $P(5, 2)$.

4.⁻ Determine the total chromatic number of the following.

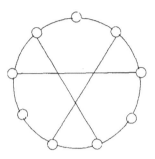

Figure 4.1

5. Suppose u and v lie on an odd cycle of F_2 and $uv \in E(F_2)$. Prove that if $|A(u) \cup A(v)| \leq 3$, then at least one of u and v is a t-vertex. (For $F_2, A(u)$ and t-vertex, we refer to the proof of Theorem 4.9.)

6. For any integer $\Delta \geq 2$, let $\mathcal{G}_{ij}^{\Delta}$ be the collection of Δ-regular graphs which are Class i and Type j, $i, j = 1, 2$. (We already know that $C_6 \in \mathcal{G}_{11}^2$, $C_4 \in \mathcal{G}_{12}^2$, $C_3 \in \mathcal{G}_{21}^2$, $C_5 \in \mathcal{G}_{22}^2$, $K_4 \in \mathcal{G}_{12}^3$ and $P(5,2) \in \mathcal{G}_{21}^3$, where $P(5,2)$ is the Petersen graph.) Let H_1 and H_2 be the two graphs given below. Prove that $H_1 \in \mathcal{G}_{11}^3$ and $H_2 \in \mathcal{G}_{22}^3$. (This problem is contributed by H. R. Hind.)

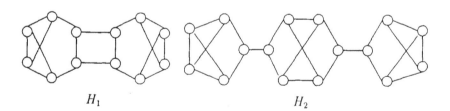

H_1 H_2

Figure 4.2

7. Applying Lemma 4.1 and Lemma 4.7, give an alternative (slightly shorter) proof of Theorem 4.9.

CHAPTER 5

GRAPHS OF HIGH DEGREE

While it was proven that the TCC holds for graphs of low degree in the 70's, it is only recently that the TCC is proved for graphs of high degree. In this chapter we shall prove the following two main results:

(i) The TCC holds for graphs G having $\Delta(G) \geq |G| - 5$.
(ii) The TCC holds for graphs G having $\Delta(G) \geq \frac{3}{4}|G|$.

The first result is due to Yap, Wang and Zhang [89], and Yap and Chew [92].

The second result is due to Hilton and Hind [93]. It is a refinement of some earlier results by Chetwynd, Hilton and Zhao Cheng [91] who showed that the TCC holds for graphs G of even order having $\delta(G) \geq \frac{3}{4}|G|$ and for graphs G of odd order having $\delta(G) \geq \frac{3}{4}(G)$ by Chew and Yap [-a]. Note that $\frac{3}{4}|G| > |G| - 5$ if and only if $|G| < 20$. Hence the second result supersedes the first result when $|G| \geq 20$.

§1. Total chromatic number of graphs G having $\Delta(G) \geq |G| - 5$

We will apply the following theorem to prove that the TCC holds for graphs G having $\Delta(G) \geq |G| - 5$.

Theorem 5.1 (Chetwynd and Hilton [85]) Let G be a connected graph of order n with three major vertices. Then G is Class 2 if and only if each of the three major vertices is of degree $n - 1$ and all the remaining vertices have degree $n - 2$.

(A proof of this lemma is reproduced in Yap [86;p.53-54].)

Theorem 5.2 (Hall, 1935) Let G be a bipartite graph having bipartition (X,Y). Then G contains a matching that saturates every vertex in X if and only if

$$|N(S)| \geq |S| \quad \text{for all} \quad S \subseteq X.$$

(A proof of this lemma can be found in many textbooks on graph theory.)

Theorem 5.3 (Chetwynd and Hilton [90]) Let G be a connected graph having four major vertices. Then G is Class 2 if and only if for some integer n, one of the following holds

(i) the valency-list of G is $(2n - 2)^{2n-3}(2n - 1)^4$;

(ii) the valency-list of G is $(2n - 2)(2n - 1)^{2n-4}(2n)^4$;

(iii) G contains a cut-edge e such that $G - e$ is the union of two graphs G_1 and G_2 where $\Delta(G_1) \le 2m - 1$ for some integer $m < n$ and the valency-list of G_2 is either $(2m - 2)(2m - 1)^{2m-4}(2m)^4$ or $(2m - 1)^{2m-2}(2m)^3$.

(Suppose n_j is the number of vertices of degree j in G. Then $1^{n_1}2^{n_2}...k^{n_k}$, where $k = \Delta(G)$, is called the valency-list of G. If $n_j = 0$ for some j, then the factor j^0 is usually omitted in the listing of the valency-list of G.)

From Theorem 5.3 it follows that if $\Delta(G) = |G| - 5$ and G has four major vertices, then G is Class 1 (This result will be applied to prove Theorem 5.6).

We say that a graph G is maximal if the minor vertices of G form a clique in G. We shall now prove the main results of this section.

Theorem 5.4 (Yap, Wang and Zhang [89]) Let G be a graph of order n having $\Delta(G) = n - k$, where $k \ge 4$ and $n \ge \max(3k - 4, 9)$. If G is maximal and $G \ge O_{k-2}$, then

$$\chi_T(G) \le \Delta(G) + 2.$$

Proof. By Lemma 2.6, we can also assume that $G \not\ge O_{k-1}$. Let $T = \{x_1, x_2, ..., x_{k-2}\}$ be a set of independent vertices in G. Let V_1 be the set of major vertices of G. Suppose $u, v \in V(G) \setminus T$ are not adjacent in G. Let M_1 be a matching in $H = G - T$ satisfying

(1) all the major vertices in $\{u, v\}$ are M_1-saturated; and

(2) $|V(M_1) \cap V_1|$ is maximum among all the matchings in H satisfying (1).

We now prove that V_1 contains at most one M_1-unsaturated vertex. Suppose otherwise. Let $z_1, z_2 \in V_1$ be M_1-unsaturated. Since $\Delta(G) = n - k$, z_1 is adjacent to at least $n - 2k + 2$ vertices $a_1, a_2, ..., a_{n-2k+2}$ in H. Clearly each a_i is M_1-saturated. Let $b_i \in V(H)$ be such that $a_i b_i \in M_1$, $i \in I = \{1, 2, ..., n - 2k + 2\}$. Clearly $b_i \ne z_2$ for any $i \in I$, but b_i can be a_j for some $j \ne i$. Suppose $z_2 b_i \in E(G)$ for some $i \in I$, then replacing $a_i b_i$ in M_1 by $\{z_1 a_i, z_2 b_i\}$ we obtain a larger matching M_1' satisfying (1) but contradicting (2). Hence $z_2 b_i \notin E(G)$ for all $i \in I$. (We call this argument as the "Enlarge-Matching Argument".) Replacing $a_i b_i$ in M_1 by $z_1 a_i$, we have $b_i \in V_1$ for each $i = 1, 2, ..., n - 2k + 2$. Since $z_1, b_1, b_2, ..., b_{n-2k+2} \notin N(z_2)$, we have $n - k = d(z_2) \leq n - (n - 2k + 2 + 2) = 2k - 4$. Hence for $k = 4$ we have a contradiction and for $k \geq 5$ any vertex of G other than $z_1, b_1, b_2, ..., b_{n-2k+2}$ is adjacent to z_2. Let $A = \{a_1, a_2, ..., a_{n-2k+2}\}$, $B = \{b_1, b_2, ..., b_{n-2k+2}\}$. Suppose $A \neq B$. We first consider the case $A \cap B \neq \phi$. By symmetry, there are vertices $c_i, c_j \in N(z_2)$ such that $c_i c_j \in M_1$. Let C be such a set of vertices. By the "Enlarge-Matching Argument", each $b_i \in B \backslash A$ is not adjacent to any $b_r \in B$ (otherwise replace $\{a_i b_i, a_r b_r\}$ by $\{z_1 a_r, b_i b_r, z_2 a_i\}$) and any $c_s \in C$. It is also not adjacent to z_1 and z_2. Hence $d(b_i) \leq n - (n - 2k + 2) - 2 - 1 = 2k - 5$. Now $2k - 5 \geq d(b_i) = n - k$ contradicts the assumption that $n \geq 3k - 4$. Hence $A \cap B = \phi$. Now since $(n - 2k + 2) + 1 \geq k - 1$ and $G[b_1, b_2, ..., b_{n-2k+2}, z_2] \not\supseteq O_{k-1}$, we have $b_i b_j \in E(G)$ for some $i, j \in I$ and by the "Enlarge-Matching Argument" we have another contradiction. Hence V_1 contains at most one M_1-unsaturated vertex, say z.

Let $G_1 = G - M_1$, $H_1 = G_1 - \{u, v\}$, and $V_2 = \{h \in V(H_1) | d_{G_1}(h) \geq \Delta - 1\}$. Since G is maximal and $G[T] = O_{k-2}$, T contains at most one minor vertex. Now let $Y = V(G) \backslash (T \cup \{u, v\})$. For any $S \subseteq T$ where S consists of only major vertices, let $s = |S|$ and $r = |N_{H_1}(S) \cap Y|$. Then by counting the number of edges joining S and $N_{H_1}(S) \cap Y$ we have $s(n - k - 2) \leq r(k - 2)$ from which it follows that $r \geq s$. Hence by Theorem 5.2, H_1 contains a matching M_2 satisfying

(3) z and all the major vertices in T are M_2-saturated; and

(4) $|V(M_2) \cap V_2|$ is maximum among all matchings in H_1 satisfying (3).

We now show that V_2 contains at most two M_2-unsaturated vertices. Suppose otherwise. Let $w_1, w_2, w_3 \in V_2$ be M_2-unsaturated. Then w_1 is adjacent to at least $n - k - 3$ vertices $a_1, a_2, ..., a_{n-k-3}$ in H_1. Clearly each a_j is M_2-saturated. Let $b_j \in V(G)$ be such that $a_j b_j \in M_2$, $j \in J = \{1, 2, ..., n - k - 3\}$. By the "Enlarge-Matching Argument", $w_2 b_j, w_3 b_j \notin E(G_1)$ for all $j \in J$. Thus $n - k - 1 \leq d_{G_1}(w_i) \leq n - (n - k - 3 + 3) = k$ for each $i = 2, 3$. Since $n \geq \max(3k - 4, 9)$ and $k \geq 4$, we now have either $k = 4$ and $n = 9$ or $k = 5$ and $n = 3k - 4 = 11$. Finally, for $n = 11$, $5 = d_{G_1}(w_i)(i = 2, 3)$ implies that $w_i \in N(a_j)$ for all $i = 2, 3$, $j = 1, 2, 3$. (If $a_j \notin N(w_i)$, then $|G| > 11$, false.) Hence $\{a_1, a_2, a_3\} \cap \{b_1, b_2, b_3\} = \phi$. If $b_i b_j \in E(G_1)$ for some $i \neq j$, then replacing $\{a_i b_i, a_j b_j\}$ in M_2 by $\{w_1 a_i, b_i b_j, w_2 a_j\}$ we obtain a larger matching M_2' satisfying (3) but contradicting (4). Hence $b_i b_j \notin E(G_1)$ for all i, j. Since w_i, $b_i \in V_2$ for all $i = 1, 2, 3$, it follows that w_i, $b_i \in N_{G_1}(u) \cap N_{G_1}(v)$ for all $i = 1, 2, 3$. Hence $d_{G_1}(u), d_{G_1}(v) \geq 6$ and so either $d(u) \geq 7$ or $d(v) \geq 7$ (because either u or v is a major vertex), which contradicts the fact that $\Delta(G) = 11 - 5 = 6$. The case $k = 4$ and $n = 9$ can be settled in the same way: We now cannot have $a_1, b_1 (= a_2) \in N(w_1)$, otherwise we would have $|G| \geq 11$, which is false. Hence we have $a_1 b_1, a_2 b_2 \in M_2$, where $\{b_1, b_2\} \cap \{a_1, a_2\} = \phi$ and by the "Enlarge-Matching Argument", $b_1 b_2 \notin E(G_1)$. Since w_1, w_2, w_3, b_1, $b_2 \in V_2$, we have w_1, w_2, w_3, b_1, $b_2 \in N_{G_1}(u) \cap N_{G_1}(v)$ from which it follows that either $d(u) \geq 6$ or $d(v) \geq 6$, a contradiction also. Hence V_2 contains at most two M_2-unsaturated vertices, say w and w'.

Finally let G^* be the graph obtained from $G_2 = G_1 - M_2$ by adjoining a new vertex v^* and adding an edge joining v^* to each vertex in $V(G) \setminus (T \cup \{u, v\})$. Then $\Delta(G^*) \leq n - k$ and G^* has at most four major vertices if $\Delta(G^*) = n - k$. By Theorem 5.1 and Theorem 5.3, or by Vizing's theorem, $\chi'(G) \leq n - k$ and consequently $\chi_T(G) \leq n - k + 2$. //

Theorem 5.5 (Yap, Wang and Zhang [89]) For any graph G of order n having $\Delta(G) \geq n - 4$,

$$\chi_T(G) \leq \Delta(G) + 2.$$

Proof. By Theorem 3.1, the TCC holds for graphs G having $\Delta(G) = |G| - 1$. Applying Lemma 2.6, it is extremely easy to show that the TCC holds for graphs G having $\Delta(G) = |G| - k$, $k = 2, 3$ (See Exericse 5(1)). By Lemma 2.1 we can assume that G is maximal. By Theorem 4.9 we can assume that $|G| \geq 9$. Finally if $\Delta(G) = |G| - 4$, then clearly $G \geq O_2 = O_{4-2}$. Hence by Theorem 5.4, the TCC holds for graphs G having $\Delta(G) = |G| - 4$. //

Theorem 5.6 (Yap and Chew [92]) Let G be a graph of order n having $\Delta(G) = n - k$, where $k \geq 5$ and $n \geq \max(13, 3k - 3)$. If G is maximal and $G \geq O_{k-3}$, then $\chi_T(G) \leq \Delta(G) + 2$.

Proof. Let $\Delta = \Delta(G)$. By Theorem 5.4, we also assume that $G \not\geq O_{k-2}$. Let

$T = \{v_1, v_2, ..., v_{k-3}\}$ be a set of independent vertices in G. Let V_1 be the set of major vertices of G. Since $G \not\supseteq O_{k-2}$ and $\delta(\bar{G}) = n - 1 - \Delta = k - 1$, we can find two pairs of nonadjacent vertices $\{x_i, y_i\}$, $i = 1, 2$, disjoint from T. Since G is maximal, we may assume that $x_1, x_2, v_1, ..., v_{k-4}$ are major vertices. We first show that $H = G - T$ contains a matching M_1 satisfying

(5) all the major vertices in $\{x_1, y_1, x_2, y_2\}$ are M_1-saturated;

(6) $|V(M_1) \cap V_1|$ is maximum among all the matchings in H satisfying (5).

We know that each major vertex of G is of degree $n - k$. Hence each major vertex in $S = \{x_1, y_1, x_2, y_2\}$ is adjacent to at least $(n-k)-(k-3)-2 = n-2k+1 \geq 4$ vertices in $V(G)\backslash(T \cup S)$. Hence, by Theorem 5.2, H contains a matching M_1 satisfying (5).

We now prove that $V_1 \backslash T$ contains at most one M_1-unsaturated vertex. Suppose otherwise. Let $u_1, u_2 \in V_1 \backslash T$ be two M_1-unsaturated vertices. As in the proof of Theorem 5.4, u_1 is adjacent to at least $n - 2k + 3$ vertices in H, whence $d(u_2) \leq n - (n - 2k + 3 + 2) = 2k - 5$, a contradiction. Hence $V_1 \backslash T$ contains at most one M_1-unsaturated vertex, say u, in H.

Let $G_1 = G - M_1$, $H_1 = G_1 - \{x_1, y_1\}$ and $V_2 = \{h \in V(H_1) | d_{G_1}(h) \geq \Delta - 1\}$. We now show that H_1 contains a matching M_2 satisfying

(7) u and all the major vertices in $T \cup \{x_2, y_2\}$ are M_2-saturated;

(8) $|V(M_2) \cap V_2|$ is maximum among all the matchings in H_1 satisfying (7).

We know that $G[v_1, v_2, ..., v_{k-3}, x_2, y_2] \not\supseteq O_{k-2}$. Hence ther exist integers $r \neq s$ such that $x_2 v_r, y_2 v_s \in E(G)$. Now each major vertex in $T \backslash \{v_r, v_s\}$ is adjacent to at least $n - k - 4 \geq 2k - 7 \geq k - 2$ vertices in $V(G)\backslash(T \cup S)$. By Theorem 5.2, there exists a matching M_2' that saturates all the major vertices in $T \backslash \{v_r, v_s\}$ in the bipartite graph having bipartition $T \backslash \{v_r, v_s\}$ and $V(G)\backslash(T \cup S)$. If u is adjacent to some $v_t \in T \backslash \{v_r, v_s\}$ and v_t is incident with an edge e of M_2', we put $M_2 = (M_2'\backslash\{e\}) \cup \{uv_t\}$. Otherwise since $d(u) = \Delta$ and $\Delta - (k - 5) - 6 \geq k - 4 \geq 1$, u is adjacent to some $u' \in V(G)\backslash(V(M_2') \cup S \cup \{v_r, v_s\})$ and we can take $M_2 = M_2' \cup \{uu'\}$.

We now show that V_2 contains at most one M_2-unsaturated vertex. Suppose otherwise. Let $z_1, z_2 \in V_2$ be M_2-unsaturated. Now z_1 is adjacent to at least $\Delta - 3$ vertices in H_1. Thus $\Delta - 1 \leq d_{G_1}(z_2) \leq n - (\Delta - 3 + 2) = k + 1$, from which it follows that $n \leq 2k + 2$, a contradiction. Hence V_2 contains at most one M_2-unsaturated vertex, say z.

Let $G_2 = G_1 - M_2$, $H_2 = G_2 - \{x_2, y_2\}$ and $V_3 = \{h \in V(H_2) | d_{G_2}(h) \geq \Delta - 2\}$. We shall show that H_2 contains a matching M_3 satisfying

(9) every vertex in $\{v_1, v_2, ..., v_{k-3}, u, z, x_1, y_1\}$ whose degree in G_2 is $\Delta - 1$ is M_3-saturated;

(10) $|V(M_3) \cap V_3|$ is maximum among all the matchings in H_2 satisfying (9).

We know that $G[v_1, v_2, ..., v_{k-3}, x_1, y_1] \not\supseteq O_{k-2}$. Hence there exist integers $r' \neq s'$ such that $x_1 v_{r'}', y_1 v_{s'} \in E(G)$. Now, each major vertex in $T \backslash \{v_{r'}, v_{s'}\}$ is adjacent to at least $(n - k - 1) - 4 \geq 2k - 8 \geq k - 3$ vertices in $V(G) \backslash (T \cup S)$. Note that $d_{G_2}(u) = \Delta - 1$ and $d_{G_2}(z) = \Delta - 1$. Hence, as before and by Theorem 5.2, H_2 contains a matching M_3 satisfying (9).

We next show that V_3 contains at most one M_3-unsaturated vertex. Suppose otherwise. Let $w_1, w_2 \in V_3$ be M_3-unsaturated. As before, w_1 is adjacent to at least $\Delta - 4$ vertices $a_1, a_2, ..., a_{\Delta-4}$ in H_2 and each a_i is M_3-saturated. Let $b_i \in V(H_2)$ be such that $a_i b_i \in M_3$, $i = 1, 2, ..., \Delta - 4$. By the "Enlarge Matching Argument", $b_i w_2 \notin E(G_2)$ for all $i = 1, 2, ..., \Delta - 4$. Thus $d_{G_2}(w_2) \leq n - (\Delta - 4) - 2 = k + 2$. Now $2k - 5 \leq n - k - 2 = \Delta - 2 \leq d_{G_2}(w_2) \leq k + 2$ implies that $k \leq 7$ and $n \leq 2k + 4$.

We consider three cases separately.

Case 1. $b_i \notin A = \{a_1, a_2, ..., a_{\Delta-4}\}$ for $i = 1, 2, ..., \Delta - 4$.

We first note that if $k = 7$, then from the inequality $2k - 5 \leq \Delta - 2 \leq d_{G_2}(w_2) \leq k + 2$, it follows that $d_{G_2}(w_2) = \Delta - 2$ and thus w_2 is adjacent to every vertex in A; also if $k = 6$ or $k = 5$ (and thus $n \geq 13$), then w_2 is not adjacent to at most one vertex in A. Hence if $b_i b_j \in E(G_2)$ for some $1 \leq i < j \leq \Delta - 4$, then replacing $\{a_i b_i, a_j b_j\}$ in M_3 by $\{w_2 a_i, w_1 a_j, b_i b_j\}$ (or $\{w_2 a_j, w_1 a_i, b_i b_j\}$), we obtain a larger matching M_3' satisfying (9) but contradicting (10). Hence $G_2[w_1, w_2, b_1, b_2, ..., b_{\Delta-4}] = O_{\Delta-2}$. Since $(M_1 \cup M_2) \cap E(G[w_1, w_2, b_1, b_2, ..., b_{\Delta-4}])$ is a union of paths and even cycles, and $\Delta - 2 = n - k - 2 \geq 2k - 5$, we have $G[w_1, w_2, b_1, b_2, ..., b_{\Delta-4}] \geq O_{k-2}$, a contradiction to the assumption.

Case 2. $b_i \in A$ and $b_j \notin A$ for some $1 \leq i, j \leq \Delta - 4$.

Suppose $b_1, b_2 \in A$, that is, $b_1 = a_2$ and $b_2 = a_1$, and suppose $b_{\Delta-4} \notin A$. By symmetry, we also have $C = \{c_1, c_2, ..., c_{\Delta-r}\} \subseteq N_{H_2}(w_2)$, $\{c_1 c_2, c_3 d_3, ..., c_{\Delta-4} d_{\Delta-4}\} \subseteq M_3$ and $d_{\Delta-4} \notin C$. First suppose $c_{\Delta-4} = a_{\Delta-4}$, say. Then $d_{\Delta-4} = b_{\Delta-4}$ and by the "Enlarge Matching Argument", $w_1, w_2, a_1, a_2, c_1, c_2, b_3, ..., b_{\Delta-5}, d_3, ..., d_{\Delta-5} \notin N_{H_2}(b_{\Delta-4})$. (Note that b_i may be equal to d_j for some i and j such that $3 \leq i \leq j \leq \Delta - 5$.) Hence

$$2k - 5 \leq n - k - 2 = \Delta - 2 \leq d_{G_2}(b_{\Delta-4}) \leq n - (\Delta - 5) - 5 = k,$$

from which it follows that $k = 5$ and $n \leq 12$, a contradiction. Next, suppose $C \cap A = \emptyset$. Then the minimum order of G is attained when $b_{\Delta-4}$ is the only vertex in $B = \{b_1, b_2, ..., b_{\Delta-4}\}$ such that $b_{\Delta-4} \notin A$. Thus $2(\Delta - 5) + 8 \leq n$ (note that $|\{x_2, y_2, w_1, w_2, a_{\Delta-4}, b_{\Delta-4}, c_{\Delta-4}, d_{\Delta-4}\}| = 8$), from which it follows that $k = 5$ and $n = 12$, again a contradiction.

Case 3. $b_j \in A$ for $j = 1, 2, ..., \Delta - 4$.

In this case $\Delta - 4$ is even. We consider the following subcases separately.

(i) $k = 7$. From $18 = 3k - 3 \leq n \leq 2k + 4 = 18$ we have $n = 18$. However, we now also have $\Delta - 4 = n - k - 4 = n - 11 = 7$ is odd, which is a contradiction.

(ii) $5 \leq k \leq 6$. By the "Enlarge Matching Argument", $d_{G_2}(a_i) \geq n - k - 2$, $i = 1, 2, ..., \Delta - 4$, and every a_i is not adjacent to any $v \in V(G_2) \backslash (\{w_1, x_2, y_2\} \cup A)$. Hence $G_2[w_1, a_2, a_2, ..., a_{\Delta-4}] = K_{\Delta-3}$. Similarly, we have $G_2[w_2, c_1, c_2, ..., c_{\Delta-4}] = K_{\Delta-3}$. Thus each vertex in $W = \{w_1, w_2\} \cup A \cup C$ must be adjacent to both x_2 and y_2, and consequently $2\Delta - 6 \leq d_{G_2}(x_2) \leq \Delta - 2$ from which it follows that $\Delta - 4$, a contradiction. Hence V_3 has at most one M_3-unsaturated vertex.

Finally, let G^* be obtained from $G_3 = G_2 - M_3$ by adding a new vertex c^* and adding an edge joining c^* to each vertex in $V(G_3) \backslash (T \cup \{x_1, x_2, y_1, y_2\})$. Then $\Delta(G^*) = \Delta - 1$, and by the choices of M_1, M_2 and M_3, G^* has at most four major vertices. Hence, by Theorem 5.1 and Theorem 5.3, $\chi_1(G^*) = \Delta - 1$ and if π is a $(\Delta - 1)$-edge-colouring of G^*, then we can modify π to a total-colouring φ of G using $(\Delta - 1) + 3$ colours. Hence $\chi_T(G) \leq (\Delta - 1) + 3 = \Delta + 2$.

The proof of Theorem 5.6 is complete. $//$

We shall further require the following lemmas.

Lemma 5.7 Let G be a garph of order 10 having $\Delta(G) = 5$. If G is maximal and $G \geq O_3$, then $\chi_T(G) \leq \Delta(G) + 2$.

Proof. The proof is almost identical to the proof of Theorem 5.4. Hence we need only to modify some parts of the proof of Theorem 5.4.

In the case of showing that V_1 contains at most one M_1-unsaturated vertex, we let $\{x\} = V(G)\backslash(T \cup \{z_1, z_2, a_1, a_2, b_1, b_2\})$ if $b_i \notin A = \{a_1, a_2\}$. Then clearly $xz_2 \notin E(G)$. Consequently, by the "Enlarge Matching Argument", $a_1, a_2 \notin N(z_2)$, a contradiction. On the other hand, if $b_i \in A$, we let $c_1, c_2 \in N_H(z_2)$. Then by the "Enlarge Matching Argument", a_1, a_2, c_1, c_2, z_1 and z_2 are all adjacent to each vertex in T, which is false. Hence V_1 contains at most one M_1-unsaturated vertex, z say.

In the case of showing that V_2 contains at most two M_2-unsaturated vertices, we now have $n - k - 1 \leq d_{G_1}(w_i) \leq k$ for each $i = 2, 3$.

Again, we let $\{x\} = V(G)\backslash\{u, v, w_1, w_2, w_3, a_1, a_2, b_1, b_2\}$. Then x is M_3-unsaturated and $xw_i \notin E(G_1)$ for all $i = 1, 2, 3$. By the "Enlarge Matching Argument", $b_1, b_2 \in V_2$, and $a_1, a_2 \in N_H(w_j)$, $j = 2, 3$. Hence $xb_1, xb_2, b_1b_2 \notin E(G_1)$. Thus $b_1, b_2, w_1, w_2, w_3 \in N(u) \cap N(v)$ and consequently either $d(u) \geq 6$ or $d(v) \geq 6$, a contradiction. //

Lemma 5.8 Let G be a graph of order $n = 10, 11$ or 12. Suppose $\Delta(G) = n - 5$, G is maximal and $G \ngeq O_3$. Then $\chi_T(G) \leq \Delta + 2$.

Proof. We shall apply the technique used in the proof of Theorem 5.6 by explicitly constructing out the three matchings M_1, M_2 and M_3.

Let $u_1 \in V(G)$ be such that $d(u_1) = \delta = \delta(G)$ and let $v \in V(G)$ be such that $u_1v \notin E(G)$. Thus v is a major vertex. Let $N_{\bar{G}}(v) = \{u_1, u_2, u_3, u_4\}$ and $N_G(v) = \{v_1, v_2, ..., v_{n-5}\}$. Since $G \geq O_3$, $G[u_1, u_2, u_3, u_4] = K_4$. If $d(u_1) = 3$, then $N_{\bar{G}}(u_1) = \{v, v_1, v_2, ..., v_{n-5}\}$ forms a clique in G, that is, $G = K_4 \cup K_{n-4}$. Thus $\chi_T(G) \leq \Delta(G) + 2$. Hence we assume that $d(u_1) = \delta \geq 4$. Let $v_1 \in N(u_1)$ and let v_2 be such that $v_1v_2 \notin E(G)$. Since $\delta \geq 4$, we may also assume that $v_{n-6}, v_{n-5} \notin N(u_1)$ (note that there is also another vertex v_j, $2 \leq j \leq n - 7$, such that $v_j \notin N(u_1)$ but we shall not specifically assume which one.) Thus $G[v, v_{n-6}, v_{n-5}] = K_3$.

We now conisder the following cases separately.

Case 1. $n = 10$ or 11.

(i) Suppose $v_3 \notin N(v_{n-6}) \cap N(v_{n-5})$, say $v_3 \notin N(v_{n-6})$.

For $n = 10$, $G - \{v_1, v_2\}$ has a 1-factor $M_1 = \{u_1 u_2, u_3 u_4, v_3 v, v_4 v_5\}$, $G - M_1 - \{v_3, v_4\}$ has a 1-factor $M_2 = \{u_1 u_4, u_2 u_3\} \cup FG[v_1, v_2, v, v_5]$ where $FG[v_1, v_2, v, v_5]$ denotes a 1-factor in $G[v_1, v_2, v, v_5]$, and we can choose M_3 to be any maximum matching in $G - (M_1 \cup M_2) - \{u_1, v\}$.

For $n = 11$, we have $G[v_4, v_1, v_2] \neq O_3$, $G[v_4, v_3, v_5] \neq O_3$ and $G[v_4, u_1, v_6] \neq O_3$. Hence by symmetry, we can assume that $v_4 u_4 \notin E(G)$. Now $G - \{v_1, v_2\}$ has a matching $M_1 = \{v_3 u_1, v_4 v, v_5 v_6, u_3 u_4\}$ which misses u_2, $G - M_1 - \{v_3, v_5\}$ has a matching $M_2 = \{u_1 u_4, u_2 u_3\} \cup FG[v, v_1, v_2, v_4]$ which misses v_6, and $G - (M_1 \cup M_2) - \{u_4, v_4\}$ has a matching $M_3 = \{u_1 u_2, v v_6\} \cup FG[v_1, v_2, v_3, v_5]$ which misses u_3.

(ii) $v_3 v_{n-6}, v_3 v_{n-5} \in E(G)$.

By symmetry, we can also assume that $v_4 v_{n-6}, v_4 v_{n-5} \in E(G)$ if $n = 11$. Thus, by symmetry again, we may further assume that $v_{n-6} u_3, v_{n-5} u_4 \notin E(G)$.

For $n = 10$, $G - \{v_1, v_2\}$ has a 1-factor $M_1 = \{u_1 u_2, u_3 u_4, v_3 v, v_4 v_5\}$, $G - M_1 - \{u_1, v_5\}$ has a 1-factor $M_2 = \{v_3 v_4, u_2 u_4\} \cup FG(v_1, v_2, v, u_3]$, and we can choose M_3 to be any maximum matching in $G - (M_1 \cup M_2) - \{u_4, v\}$.

For $n = 11$, $G - \{v_1, v_2\}$ has a matching $M_1 = \{u_1 u_4, u_2 u_3, v_3 v, v_5 v_6\}$ which misses v_4, $G - M_1 = \{u_3, v_5\}$ has a matching $M_2 = \{u_2 u_4, u_3 u_6\} \cup FG[v_1, v_2, v_4, v]$ which misses u_1, and $G - (M_1 \cup M_2) - \{u_4, v_6\}$ has a matching $M_3 = \{v_4 v, u_1 u_2\} \cup FG[v_1, v_2, v_5, u_3]$ which misses v_3.

Case 2. $n = 12$.

This case is very similar to Case 1. Although the length of proof of Case 2 is slightly shorter than that of Case 1, the proof is very technical and thus we omit it.
//

Finally, from Theorem 5.6, Lemma 5.7 and Lemma 5.8, we deduce

Theorem 5.9 (Yap and Chew [92]) For any graph G of order n having $\Delta(G) = n-5$,

$$\chi_T(G) \leq \Delta(G) + 2.$$

Proof. By Lemma 2.1, we can assume that G is maximal. By Theorem 5.4, if $n \geq 11$ and $G \geq O_3$, then this theorem is true. By Theorem 5.6, if $n \geq 13$, then this theorem is also true. Since the TCC holds for graphs G having $\Delta(G) \leq 4$, this theorem is true for $n \leq 9$. Hence we need only to consider the following remaining cases:

(i) $n = 10$ and $G \geq O_3$;

(ii) $n = 10, 11$ or 12 and $G \not\geq O_3$.

However, these two cases have been settled by Lemma 5.7 and Lemma 5.8. //

Remark. From the proofs of Theorem 5.6 and Theorem 5.9, we see that the technique we apply will be extremely difficult to prove that the TCC also holds for graphs of order n having $\Delta(G) = n - 6$.

§2. Total chromatic number of graphs G having $\Delta(G) \geq \frac{3}{4}|G|$

Before we prove the main results of this section, we define a few terms:

Let φ be a total-colouring of a graph G and let α and β be two distinct colours. We call a path $P = x_0 e_1 x_1 e_2 \ldots e_n x_n$ an $\underline{(\alpha,\beta)\text{-path}}$ if the edges $e_i = x_{i-1}x_i$ are alternately coloured α and β, or β and α.

An (α,β)-path $P = x_0 e_1 x_1 e_2 \ldots e_n x_n$ is $\underline{\text{maximal}}$ if

(a) x_0 is incident with no edge other than e_1 coloured α or β and

(b) x_n is incident with no edge other than e_n coloured α or β.

If, in addition to (a) and (b), the (α,β)-path P satisfies

(c) neither x_0 nor x_n is coloured α or β, then P is said to be $\underline{\text{swoppable}}$; otherwise

P is said to be $\underline{\text{blocked}}$ or $\underline{\text{unswoppable}}$.

The $\underline{\text{reverse}}$ of a path $P = x_0 e_1 x_1 e_2 \ldots e_n x_n$ is the path $x_n e_n x_{n-1} \ldots e_1 x_0$ and is denoted by P^{-1}.

Let G be a graph and let η be a function from $E(G)$ to $K = \{1, 2, ..., \Delta(G) + 2\}$. Let $P(K)$ be the power set of K. We call a function

$$\varphi : V(G) \cup E(G) \to P(K)$$

a (proper) total-set-colouring of G (with respect to η) using $\Delta(G) + 2$ colours if

(i) $|\varphi(v)| = 1$ for all $v \in V(G)$.

(ii) $1 \leq |\varphi(e)| \leq \eta(e)$ for all $e \in E(G)$.

(iii) $\varphi(a) \cap \varphi(b) = \phi$ if a and b are incident or adjacent.

Suppose φ is a total-set-colouring of G using $\Delta(G) + 2$ colours $1, 2, ..., \Delta(G) + 2$. For any $v \in V(G)$ we use $m_\varphi(v)$ to denote the set of colours in $\{1, 2, ..., \Delta(G) + 2\}$ absent at v, i.e. the set of colours not assigned by φ to the vertex v or to an edge incident with v.

Suppose $uv \in E(G)$ and φ is a total-set-colouring of $G - uv$. We define a fan at u to be a pair $((u_0 = u, u_1, ..., u_r), (\beta_1, \beta_2, ..., \beta_r))$ of sequences of distinct vertices and colours such that for $1 \leq i \leq r$, $\beta_i \in m_\varphi(u_{i-1})$ and u_i is the (unique) vertex for which $\beta_i \in \varphi(uu_i)$.

We shall apply the following three results to prove that the TCC holds for graphs G having $\Delta(G) \geq \frac{3}{4}(G)$ in this section.

Theorem 5.10 (Flandrin, Jung and Li [91]) Let G be a 2-connected graph. If, for each set $\{v_1, v_2, v_3\}$ of three independent vertices,

$$d(v_1) + d(v_2) + d(v_3) \geq |G| + |N(v_1) \cap N(v_2) \cap N(v_3)|,$$

then G is hamiltonian.

(Note that if G is 2-connected and $\bar{G} \not\supseteq O_3$ then by a well-known result of V. Chvátal and P. Erdös which says that for any graph G of order at least 3, if $\kappa(G) \geq \alpha(G)$, then G is hamiltonian, G is always hamiltonian.)

Theorem 5.11 (Hajnal and Szemerédi [70]) If G is a graph and r is an integer such that $r > \Delta(G)$, then G has a vertex-colouring φ using r colours $1, 2, ..., r$ such that for any two distinct colour classes $\varphi^{-1}(i)$ and $\varphi^{-1}(j)$,

$$\left||\varphi^{-1}(i)| - |\varphi^{-1}(j)|\right| \leq 1.$$

(A proof of this result can also be found in Chapter 6 of B. Bollobás : Extremal Graph Theory, Academic Press, London, 1978.)

Theorem 5.12 (Dirac, 1952) If G is a graph of order $p \geq 3$ and $\delta(G) \geq \frac{p}{2}$, then G is hamiltonian.

We first prove the main result (Theorem 5.13) for the case that G is of odd order. The case that G is of even order (Theorem 5.14) can be proved easily using Theorem 5.13.

Theorem 5.13 (Hilton and Hind [93]) If G is a graph having odd order and $\Delta(G) \geq \frac{3}{4}|G| - 1$, then

$$\chi_T(G) \leq \Delta(G) + 2.$$

Proof. By Lemma 2.1, we can assume that G is maximal. Let $n = |G|$, $\Delta = \Delta(G)$ and $V = V(G)$.

Since the TCC holds for graphs G having $\Delta(G) \leq 4$ and $\Delta(G) \geq n - 5$, we may assume that $n \geq 9$. Let $k = n - \Delta - 2$. Then $k \leq \Delta - \lceil \frac{n}{2} \rceil$. By Theorem 5.11, G has a vertex-colouring φ using $\Delta + 2$ colours $1, 2, ..., \Delta + 2$ such that each colour class has cardinality one or two. Suppose

$$|\varphi^{-1}(i)| = 2 \text{ for } i = 1, 2, ..., k \text{ and } |\varphi^{-1}(i)| = 1 \ \ i = k + 1, k + 2, ..., \Delta + 2.$$

Let $\varphi^{-1}(i) = \{x_i, y_i\}$ for $i = 1, 2, ..., k$. Then $x_i y_i \notin E(G)$.

We now define a sequence of graphs $G_0 = G, G_1, ..., G_k$ such that for each $i \leq k$ the vertices of degree less than $\Delta - i$ in G_i induce a complete graph. While defining each G_i for $i = 1, 2, ..., k$, we also define two matchings F_i and F_i^*, and a vertex w_i.

For $i = 1, 2, ..., k$, let $H_i = G_{i-1} - \{x_i, y_i\}$. When defining F_i (a near perfect matching of G) and w_i, we consider three cases separately:

Case 1. H_i is 2-connected.

From the fact that all the vertices of degree less than $\Delta - (i - 1)$ in G_{i-1} are adjacent in G_{i-1}, it follows that in any set of three independent vertices in H_i, at most one vertex can have degree less than $\Delta - (i - 1) - 2 = \Delta - i - 1$. Thus if

$\{v_1, v_2, v_3\}$ is a set of independent vertices in H_i, then

$$\sum_{j=1}^{3} d_{H_i}(v_j) \geq 2(\Delta - i - 1) + \min_{j=1,2,3}\{d_{H_i}(v_j)\}$$

$$\geq 2(\Delta - i - 1) + |\bigcap_{j=1}^{3} N_{H_i}(v_j)| \geq (n - 2) + |\bigcap_{j=1}^{3} N_{H_i}(v_j)|,$$

since $i \leq k \leq \Delta - \lceil \frac{n}{2} \rceil$. Since H_i is 2-connected, it follows from Theorem 5.10 that H_i is hamiltonian. Thus H_i has a near perfect matching F_i which avoids a vertex w_i.

Case 2. H_i is 1-connected, but not 2-connected.

Let u_i be a cut-vertex of H_i and let $v \in V(H_i) \setminus \{u_i\}$. Suppose $d_{G_{i-1}}(v) \geq \Delta - (i - 1)$. Then since $i \leq k = n - \Delta - 2 \leq \frac{n}{4} - 1$, we have

$$d_{H_i}(v) \geq \Delta - (k - 1) - 2 \geq \left(\frac{3n}{4} - 1\right) - \left(\frac{n}{4} - 2\right) - 2 = \frac{n}{2} - 1.$$

Thus all such vertices occur in a single component H_i' of $H_i - u_i$. Furthermore since the vertices $v \in V(H_i) \setminus \{u_i\}$ of degree less than $\Delta - (i - 1)$ induce a complete subgraph, they too form a single component H_i'' of $H_i \setminus \{u_i\}$. Since $|H_i''| \geq 1$, we have $|H_i'| \leq n - 2$. Hence for each $v \in V(H_i')$, $d_{H_i'}(v) \geq \frac{n}{2} - 1 = \frac{1}{2}(n - 2)$ and by Dirac's theorem, H_i' is hamiltonian. Clearly H_i has a near perfect matching F_i which avoids a vertex $w_i \in V(H_i')$.

Case 3. H_i is not connected.

Using the same techniques as in Case 2 above, it can be shown that H_i has exactly two components H_i' and H_i'' such that H_i' is hamiltonian and $H_i'' \simeq K_r$. Since $|H_i|$ is odd, the vertex w_i is a vertex in the odd component of H_i.

Let

$$F_i^* = \{uv \in F_i | \max\{d_{G_{i-1}}(u), d_{G_{i-1}}(v)\} \geq \Delta - (i - 1)\}.$$

Let $G_i = G_{i-1} - F_i^*$. It is necessary to note that if $d_{G_{i-1}}(u)$ and $d_{G_{i-1}(v)}$ are both less than $\Delta - (i - 1)$ then by assumption that G_0 is maximal, uv must be an edge in G_0 and by the constructions of $G_1, G_2, ..., G_i$, $uv \in E(G_j)$ for all $j = 1, 2, ..., i$.

We define a function $\eta_0 : E(G) \to \{1, 2, ...\}$ by

$$\eta_0(e) = 1 \text{ if } e \notin \bigcup_{j=1}^{k} F_j; \text{ and } \eta_0(e) = |\{F_j | e \in F_j\}| \text{ if } e \in \bigcup_{j=1}^{k} F_j.$$

48

We say that a total-set-colouring, ψ of G with respect to η_0, is <u>suitable</u> if:

(11) for each $i \in \{1, 2, ..., k\}$,

    (i) $V \cap \psi^{-1}(i) = \{x_i, y_i\}$, and

    (ii) there is exactly one vertex in G in which colour i is missing;

(12) for each $i \in \{k+1, k+2, ..., \Delta+2\}$, $|V \cap \psi^{-1}(i)| = 1$.

It is not difficult to prove that for each vertex v in G the sum of the values $\eta_0(e)$, summed over the edges e incident with v, is at most Δ (see Exercise 5(5)). Let E^* be a maximal subset of $E(G)$ for which we can find a suitable total-set-colouring π of $G' = G(V, E^*)$, the spanning subgraph of G having edge-set E^*. (Clearly such an E^* exists when we take $E^* \supseteq F_1 \cup ... \cup F_k$.)

If $E^* = E(G)$, there is nothing more to prove. Hence we assume that $E^* \neq E(G)$. Let $e_0 = u_0 v_0 \in E(G) \backslash E^*$ and let $H = G(V, E^* \cup \{e_0\})$. We shall show that π can be modified to yield a suitable total-set-colouring π' of H using the same set of colours $\{1, 2, ..., \Delta+2\}$ which contradicts the assumption.

We begin by noting that if $((v_0, v_1, ..., v_r), (\beta_1, ..., \beta_r))$ is a fan at u_0 then $|\pi(u_0 v_i)| = 1$ for all $i = 1, 2, ..., r$. For suppose that j is the smallest index for which $|\pi(u_0 u_j)| \geq 2$, then π can be modified to a required π' by putting

$$\pi'(u_0 v_0 = \{\beta_1\},$$

$$\pi'(u_0 v_{i-1}) = (\pi(u_0 v_{i-1}) \backslash \{\beta_{i-1}\}) \cup \{\beta_i\} \text{ for } i = 2, ..., j,$$

$$\pi'(u_0 v_j) = \pi(u_0 v_j) \backslash \{\beta_j\},$$

and $\pi'(a) = \pi(a)$ for any other element a in $V \cup E(H)$.

(We call this as the "<u>fan recolouring process</u>".)

Next we show that we may assume that one of the colours $k+1, k+2, ..., \Delta+2$ is missing at u_0. If some colours from $\{k+1, k+2, ..., \Delta+2\}$ are missing at v_0, we can interchange the labels of u_0 and v_0. Hence we consider the case that none of the colours $k+1, k+2, ..., \Delta+2$ is missing at either u_0 or v_0.

If $\beta_j \in \{k+1, k+2, ..., \Delta+2\}$ for some $1 \leq j \leq r$, then we can modify π to a required π' by setting $\pi'(u_0 v_{i-1}) = \pi(u_0 v_i)$ for $i = 1, 2..., j$, and $\pi'(a) = \pi(a)$ for any other element a in $V \cup E(H)$, and relabeling u_0 as v_0 and v_j as u_0. The new suitable total-set-colouring π' of H has the property that one of the colours $k+1, k+2, ..., \Delta+2$ is missing at u_0. Thus we suppose that $m_\pi(u_0) \cup m_\pi(v_0) \cup \{\beta_1, ..., \beta_r\} \subseteq \{1, 2, ..., k\}$.

Since the sum of the values $\eta_0(e)$, summed over the edges incident with v_0 is at most Δ, applying the "fan recolouring process" if necessary, we know that there are at most $\Delta - 1$ colours in the colour-sets assigned by π to edges incident with v_0. Hence there at least $(\Delta + 2) - ((\Delta - 1) + 1) = 2$ colours, β_1 and β_1' say, absent at v_0. Let $((v_0, v_1, ..., v_p), (\beta_1, ..., \beta_p))$ and $((v_0, v_1', ..., v_q'), (\beta_1', ..., \beta_q'))$ be two fans at u_0 for which the indices p and q are maximal. Since $\beta_1, ..., \beta_p, \beta_1', ..., \beta_q' \in \{1, 2, ..., k\}$, therefore each of these colours is in $m_\pi(x)$ for exactly one vertex x ($\beta_i \in m_\pi(v_{i-1}), \beta_j' \in m_\pi(v_{j-1}')$). Hence

$$m_\pi(v_p) \cap \{\beta_1, ..., \beta_p\} = \phi \quad \text{and} \quad m_\pi(v_q') \cap \{\beta_1', ..., \beta_q'\} = \phi.$$

Clearly $m_\pi(v_p) \cap m_\pi(u_0) = \phi$. Now since the fan is maximal, and since $m_\pi(v_p) \cap \{m_\pi(u_0) \cup \{\beta_1, ..., \beta_p\}\} = \phi$, it follows that $m_\pi(v_p) = \pi(u_0)$. Similarly $m_\pi(v_q') = \pi(u_0)$. We must consider two options:

(i) If $v_p = v_q'$, let v_j be the first vertex in the sequence $v_1, v_2, ..., v_p$ which is also in the sequence $v_1', v_2', ..., v_q'$, say $v_j = v_i'$. It follows that $\beta_j \in m_\pi(v_{j-1}) \cap m_\pi(v_{i-1}')$, so β_j is missing at two vertices, which is a contradiction.

(ii) If $v_p \neq v_q'$, then $\pi(u_0) \subseteq m_\pi(v_p) \cap m_\pi(v_q')$, so that the colour γ assigned to u_0 is missing at two vertices, i.e. $\gamma \in \{k + 1, k + 2, ..., \Delta + 2\}$. Again we can produce a new suitable total-set-colouring π' by setting $\pi'(u_0 v_{i-1}) = \pi(u_0 v_i)$ for $i = 1, 2, ..., p$, $\pi'(a) = \pi(a)$ for any other element $a \in V \cup E(H)$, and interchanging the labels v_p with u_0. The colouring π' has the required property that one colour from $\{k + 1, k + 2, ..., \Delta + 2\}$ is missing at u_0.

Thus we conclude that we may assume that one of the colours $k + 1, k + 2, ..., \Delta + 2$ is missing at u_0.

Finally we show that the existence of a suitable total-set-colouring π of $H - e_0$ with a colour from $\{k + 1, k + 2, ..., \Delta + 2\}$ missing at u_0 implies the existence of a suitable total-set-colouring of H.

Let $\alpha \in m_\pi(u_0) \cap \{k + 1, k + 2, ..., \Delta + 2\}$. Since $|m_\pi(v_0)| \geq 2$, we can choose $\beta_1 \in m_\pi(v_0)$ such that $\{\beta_1\} \neq \pi(u_0)$. Let $((v_0, v_1, ..., v_p), (\beta_1, ..., \beta_p))$ be a fan at u_0 where we terminate the sequences at v_p and β_p ($p \geq 1$) if

(13) there exists a $j \in \{1, 2, ..., p - 1\}$ such that $\beta_j \in m_\pi(v_p)$, or

(14) $\pi(u_0) \subseteq m_\pi(v_p)$.

(By the construction of a maximal fan at u_0, one of these two cases must occur.)

Case 1. Suppose there is a $j \in \{1, 2, ..., p-1\}$ such that $\beta_j \in m_\pi(v_p)$.

Now, $\beta_j \in m_\pi(v_{j-1}) \cap m_\pi(v_p)$ implies that $\beta_j \in \{k+1, k+2, ..., \Delta+2\}$.

Consider the (perhaps trivial) maximal (α, β_j)-paths P, Q and R beginning at v_{j-1}, u_0 (through v_j) and v_p respectively. First suppose that two of these paths are the same paths, say $P = Q^{-1}$. (The other two possibilities are similar.) We define a total-set-colouring π' for H by setting:

$$\pi'(e) = \pi(e)\Delta\{\alpha, \beta_j\} \quad \text{if} \quad e \in E(P) \setminus \{u_0 v_j\},$$
$$\pi'(u_0 v_{i-1}) = \pi(u_0 v_i) \text{ if } i \in \{1, 2, ..., p\} \setminus \{j\},$$
$$\pi'(u_0 v_{j-1}) = \{\alpha\}, \ \pi'(u_0 v_p) = \{\beta_j\},$$

and

$$\pi'(a) = \pi(a) \quad \text{for any other element} \quad a \in V \cup E(H).$$

(Note that the notation Δ here denotes the symmetric difference of two sets.) The total-set-colouring π' satisfies $\pi'|_V = \pi|_V$. Furthermore, each colour is assigned by π' to at least as many edges as it was by π. Thus π' is a suitable total-set-colouring of H, a contradiction.

Next suppose that all three of the maximal (α, β_j)-paths P, Q and R are distinct. Since π is a suitable colouring, condition (12) implies that $|\pi^{-1}(\alpha) \cap V(G)| = 1$ and $|\pi^{-1}(\beta_j) \cap V(G)| = 1$. Thus at least one of P, Q and R is a swoppable (α, β_j)-path.

Suppose Q is swoppable (the other two cases are similar). We define a total-set-colouring π' of H by setting

$$\pi'(e) = \pi(e)\Delta\{\alpha, \beta_j\} \quad \text{if} \quad e \in E(Q),$$
$$\pi'(u_0 v_{i-1}) = \pi(u_0 v_i) \quad \text{for} \quad i \in \{1, 2, ..., j\},$$

and

$$\pi'(a) = \pi(a) \quad \text{for any other element} \quad a \in V \cup E(H).$$

As before, it is not difficult to see that π' is a suitable total-set-colouring of H, another contradiction.

Case 2. Suppose $\pi(u_0) \subseteq m_\pi(v_p)$.

Let $\beta' \in m_\pi(v_0) \setminus \{\beta_1\}$. We first consider the case that $\{\beta_1'\} \neq \pi(u_0)$. Again let $((v_0' = v_0, v_1', ..., v_q'), (\beta_1', ..., \beta_q'))$ be a fan at u_0 where we terminate the sequences at v_q' and β_q' $(q \geq 1)$ if

(13)$'$ there exists a $j \in \{1, 2, ..., q-1\}$ such that $\beta_j' \in m_\pi(v_q')$ or

(14)$'$ $\pi(u_0) \subseteq m_\pi(v_q')$.

By the argument of Case 1, we now need only to consider the case (14)$'$. Let $\{\gamma\} = \pi(u_0)$. Then $\gamma \in m_\pi(v_p) \cap m_\pi(v_q')$. First suppose $v_q' \neq v_p$. Then $\gamma \in \{k+1, k+2, ..., \Delta+2\}$. Thus u_0 is the only vertex assigned the colour γ.

Consider the maximal (α, γ)-paths P and Q beginning at v_p and v_q' respectively. Neither P nor Q can end at u_0 since there is no edge incident with u_0 coloured α or γ. Furthermore, since $\alpha \in \{k+1, k+2, ..., \Delta+2\}$ only one vertex of G is coloured α. Thus at least one of P and Q is swoppable (even if $P = Q^{-1}$). Suppose P is swoppable. Then we can obtain a required π' for H by setting

$$\pi'(e) = \pi(e)\Delta\{\alpha, \gamma\} \quad \text{if} \quad e \in E(P),$$

$$\pi'(u_0v_{i-1}) = \pi(u_0v_i) \quad \text{if} \quad i \in \{1, 2, ..., p\},$$

$$\pi'(u_0v_p) = \{\alpha\}, \quad \text{and} \quad \pi'(a) = \pi(a) \quad \text{for any other element} \quad a \in V \cup E(H).$$

It is not difficult to see that π' is a suitable colouring of H, a contradiction.

Next, suppose $v_q' = v_p$. Let j be the smallest index such that $v_j \neq v_i'$, $i \leq q$, for which $\beta_{j+1} = \beta_{i+1}' = \beta$. Then we apply the same "fan recolouring process" to one of the three maximal (α, β)-paths beginning at v_j, u_0, and v_i'. We obtain a contradiction in the same way as what we have proved earlier.

Since in every case we arrive at a contradiction, it follows that $E^* = E(G)$ and thus $\chi_T(G) \leq \Delta + 2$. $//$

Theorem 5.14 (Hilton and Hind [93]) If G is a graph having even order and $\Delta(G) \geq \frac{3}{4}|G|$, then

$$\chi_T(G) \leq \Delta(G) + 2.$$

Proof. Add a new vertex v^* to G and turn it into a maximal graph having $\Delta(G^*) = \Delta(G)$. Then

$$\Delta(G^*) = \Delta(G) \geq \frac{3}{4}|G| = \frac{3}{4}(|G^*| - 1) \geq \frac{3}{4}|G^*| - 1.$$

Hence by Theorem 5.13, $\chi_T(G^*) \leq \Delta(G^*) + 2$ and thus $\chi_T(G) \leq \chi(G^*) \leq \Delta(G^*) + 2 = \Delta(G) + 2$. $//$

Combining Theorem 5.13 and Theorem 5.14, we have

Theorem 5.15 (Hilton and Hind [93]) If G is a graph having $\Delta(G) \geq \frac{3}{4}|G|$, then

$$\chi_T(G) \leq \Delta(G) + 2.$$

Exercise 5

1.⁻ Prove that for any graph G of order n having $\Delta(G) = n - k$, $2 \leq k \leq 3$,

$$\chi_T(G) \leq \Delta(G) + 2.$$

2. Let G be a d-regular graph of odd order $p \geq 3$. Prove that if $\chi_T(G) = d + 1$, then $\overline{G}$ contains the disjoint union $K_{p_1} \cup K_{p_2} \cup ... \cup K_{p_r}$, where each $p_i \geq 3$ is odd and $p_1 + p_2 + ... + p_r - r = p - 1 - d$.
(Chetwynd, Hilton and Zhao Cheng [91])

3. Let G be a d-regular graph of even order $p \geq 3$. Suppose $d \geq \frac{3}{4}(p - 1)$. Prove that if $\overline{G}$ does not contain a disjoint union $K_{p_1} \cup K_{p_2} \cup ... \cup K_{p_s}$, where $s \leq d + 1$, each $p_i \geq 2$ is even, and $p_1 + p_2 + ... + p_s = p$, then $\chi_T(G) = d + 2$. (Chetwynd, Hilton and Zhao Cheng [91])

4.⁻ Let G be a d-regular graph of odd order $p \geq 3$. Suppose $d \geq p - 5$. Prove that G is Type 1 if $\overline{G}$ contains K_3, and is Type 2 otherwise.

5. Let η_0 be the function defined in the proof of Theorem 5.13. Prove that for each vertex v in G, the sum of the values $\eta_0(e)$, summed over all the edges e incident with v, is at most Δ.

CHAPTER 6

CLASSIFICATION OF
TYPE 1 AND TYPE 2 GRAPHS

In Chapter 3 we gave a complete classification of the balanced complete r-partite graphs according to their total chromatic numbers. In this chapter, we shall further classify graphs of extremely high degree according to their total chromatic numbers. Obviously the first target must be graphs G of even order p having $\Delta(G) = p - 1$ (For odd p, since $\chi_T(K_p) = p = \Delta(G) + 1$, any graph G of order p having $\Delta(G) = p - 1$ must be Type 1). This pioneer work (Theorem 6.1) is due to Hilton [89]. Same classification problems which have been completely solved include:

(i) $|G| = 2n$, $\Delta(G) = 2n - 2$ (Chen and Fu [92]).

(ii) $|G| = 2n + 1$, $\Delta(G) = 2n - 1$ (Yap, Chen and Fu [-a]).

(iii) Odd order d-regular graphs G, where $d \geq \frac{\sqrt{7}}{3} |G|$ (Chetwynd, Hilton and Zhao [91]).

(iv) G is nearly complete bipartite (Hilton [91]); graphs G such that $\bar{G}$ is bipartite (Dugdale and Hilton [94]).

The above results, except (iii), will be presented in this chapter.

§1. Even order graphs G having $\Delta(G) = |G| - 1$

In this section we present Hilton's pioneer result on the classification of graphs G of high maximum degree according to their total chromatic numbers.

Theorem 6.1 (Hilton [89]) Suppose G is a graph of order $2n$ having $\Delta(G) = 2n - 1$. Then G is Type 1 if and only if

(1) $$e(\bar{G}) + \alpha'(\bar{G}) \geq n.$$

Proof. We let $e(\bar{G}) = e$ and $\alpha'(\bar{G}) = m$. By Theorem 3.1, $\chi_T(G) \leq 2n+1$. Suppose $\chi_T(G) = 2n$. Then by Theorem 2.7, $e + m \geq n$.

It remains to prove the sufficiency. By adding in edges, if necessary, we may without loss of generality, suppose that $e + m = n$ or n is odd and $\bar{G}$ consists of $\frac{1}{2}(n + 1)$ independent edges (so that $e + m = n + 1$), and then show that G has a total-colouring using $2n$ colours.

Let $\{v_1, ..., v_y\}$ be the set of vertices in $\bar{G}$ each having degree at least 1. Let $V(K_{2n}) = \{v_1, ..., v_y, v_{y+1}, ..., v_{2n}\}$, and let $M = \{e_1, e_2, ..., e_m\}$, where $e_i = v_i v_{m+i}$, be a maximum matching in $\bar{G}$.

Suppose for the moment that $e + m = n$. Let H^* be the graph obtained by adjoining a new vertex v^* to $G + M$ and joining v^* to each of $v_{2m+1}, ..., v_{2n}$. Clearly, $\Delta(H^*) = 2n$. From $2n - d_{H^*}(v^*) = 2m$ and $\sum_{i=1}^{y}(2n - d_{H^*}(v_i)) = 2e$, it follows that

$$\sum_{v \in V(H^*)}(2n - d_{H^*}(v)) = 2(e + m) = 2n.$$

We next construct a multigraph H^{**} by adjoining a new vertex v^{**} to H^* and joining v^{**} to each $v \in \{v^{**}, v^*, v_1, ..., v_{2n}\}$ by $2n - d_{H^*}(v)$ edges (so some of these edges may be multiple). Then $d_{H^{**}}(v) = 2n$ for each $v \in \{v^{**}, v^*, v_1, ..., v_{2n}\}$. Also in H^{**}, v^* and v^{**} are joined by $2m$ edges. For $y \leq x \leq 2n$, let H_x^{**} denote the subgraph of H^{**} induced by $\{v^{**}, v^*, v_1, ..., v_x\}$. Since in H^{**} each vertex of $\{v^*, v_1, ..., v_y\}$ is joined to each of the $2n - y$ vertices $\{v_{y+1}, v_{y+2}, ... v_{2n}\}$, so in H_y^{**} each vertex $v \neq v^{**}$ has degree y.

If $e + m = n + 1$ and $\bar{G}$ consists of $\frac{1}{2}(n + 1)$ independent edges, we vary the above construction of H^{**} slightly. We let v^{**} be joined to v_i, v_{m+i} ($2 \leq i \leq m$) and v^* as before, but not to v_1 or v_{m+1}. The vertices v_1 and v_{m+1} are joined by two edges e_1 and e_1' instead of by just one edge. It again follows that $d_{H^{**}}(v) = 2n$ for every $v \in \{v^{**}, v^*, v_1, ..., v_{2n}\}$.

We shall show that H^{**}, and therefore H^*, has a $2n$-edge-colouring φ with the edges of M and the edges $v^* v_{2m+1}, ..., v^* v_{2n}$ all receiving different colours. Clearly from φ we can obtain a $2n$-total-colouring of G by retaining the colours on the edges of G, colouring the vertex v_i with the colour $\varphi(v^* v_i)$ ($2m + 1 \leq i \leq 2n$) and colouring the vertices v_i and v_{m+i} with the colour $\varphi(v_i v_{m+i})$ ($1 \leq i \leq m$).

For the case $e + m = n$, we edge-colour H_y^{**} as follows. We first colour the edge $v_i v_{m+i}$ with colour c_i ($1 \leq i \leq m$). We colour the $2m$ edges joining v^* and

v^{**} with colours $c_1, c_2, ..., c_{2m}$ and the remaining $2n - 2m$ edges on v^{**} are coloured with colours $c_{2m+1}, ..., c_{2n}$. After that the remaining edges of H_y^{**} are coloured one by one, greedily: if an edge e^* is uncoloured, then at most $2(y - 1)$ colours are used on the edges incident with the end-vertices of e^*. Since $e + m = n$, $\bar{G}$ consists of m independent edges $v_1 v_{m+1}, ..., v_m v_{2m}$, and a further $e - m = n - 2m$ edges, each of which is incident with at least one of the vertices $v_1, v_2, ..., v_{2m}$, these further $e - m$ edges has at most $e - m = n - 2m$ end-vertices in $\{v_{2m+1}, ..., v_y\}$. Thus $y \leq (n - 2m) + 2m = n$. Consequently, $2(y - 1) \leq 2(n - 1) < 2n$ and there is always a colour available to colour e^* with.

For the case n is odd and $\bar{G}$ consists of $\frac{1}{2}(n + 1)$ independent edges, we have $y = n + 1$ and thus the argument above does not work. However, in this case H_y^{**} has a 1-factor F containing an edge $v^* v^{**}$ and $F \cap M = \phi$. We colour the edges of F with colour c_{2m}. Apart from this, we proceed as before; this time there are at most $2(y - 1) - 1 = 2n - 1 < 2n$ colours used on edges incident with the end-vertices of e^*, so again there is always a colour available to colour e^* with.

We shall apply the following lemma in the later part of the proof.

Lemma 6.2 For $x \geq y$, an edge-colouring of H_x^{**} with colours $c_1, c_2, ..., c_{2n}$ can be extended to an edge-colouring of H^{**} with the same colours if and only if each colour c_i is used on at least $x - n + 1$ edges of H_x^{**}.

Proof of Lemma 6.2.

Suppose that an edge-colouring φ of H_x^{**} with colours $c_1, ..., c_{2n}$ can be extended to an edge-colouring of H^{**} with the same set of colours. Then for each i, $1 \leq i \leq 2n$, the number of edges not in H_x^{**} which are coloured c_i is at most $2n - x$. Since each colour class in H^{**} consists of exactly $n + 1$ edges, at least $n + 1 - (2n - x) = x - n + 1$ edges of H_x^{**} are coloured c_i.

Conversely, suppose H_x^{**} is edge-coloured with $c_1, ..., c_{2n}$ and that for each $i = 1, 2, ..., 2n$, c_i occurs on at least $x - n + 1$ edges of H_x^{**}. We shall extend this to an edge-colouring of H_{x+1}^{**} with the same set of colours in such a way that each colour will occur on at least $(x + 1) - n + 1 = x - n + 2$ edges. Iterating this will eventually yield a required edge-colouring of H^{**}.

In order to do so we first construct a bipartite graph B as follows: The bipartition of B is $(\{v^{*\prime}, v_1', ..., v_x'\}, \{c_1', ..., c_{2n}'\})$. A vertex v' is joined in B to a vertex c' if in H_x^{**} there is no edge incident with v which is coloured c. Since each vertex of H_x^{**} (except v^{**}) has degree x, each v'-vertex in B has degree $2n - x$. Since each colour c_i is used in H_x^{**} on at least $x - n + 1$ edges, there are at least $2(x - n + 1)$ vertices in H_x^{**} which are incident with an edge coloured c_i. Hence c_i fails to be on any edges incident with at most $x + 2 - 2(x - n + 1) = 2n - x$ vertices of H_x^{**} (c_i is on an edge incident with v^{**}). Therefore in B each c'-vertex has degree at most $2n - x$.

By König's theorem (see Theorem 1.1 Yap [86;p.11]), B can be edge-coloured with $2n - x$ colours. Let α be one fixed colour in such an edge-colouring. Then α occurs on every vertex of degree $2n - x$. If an edge $c'v'$ is coloured α, we colour the edge $v_{x+1}v$ in H_{x+1}^{**} with colour c. It is not difficult to check that in this way we obtain a proper edge-colouring of H_{x+1}^{**}. In this edge-colouring, any colour c which occured on only $x - n + 1$ edges of H_x^{**} gives rise in B to a vertex c' of degree $2n - x(= (x + 2) - 2(x - n + 1))$, and so c' has an edge coloured α on it. Hence c is assigned to some edge incident with v_{x+1}. It follows that in H_{x+1}^{**}, each colour does occur on at least $x - n + 2$ edges, as required. $//$

We now resume the proof of the sufficiency in Theorem 6.1.

In the case when $e + m = n$, we have $y \leq n$. Since each colour occurs on an edge of H_y^{**} incident with v^{**} and $1 \geq y - n + 1$, by Lemma 6.2, H^{**} can be edge-coloured with $2n$ colours.

In the case when $e + m = n + 1$ and $\bar{G}$ consists of $\frac{1}{2}(n + 1)$ independent edges, we have $y = n + 1$, and so the condition of Lemma 6.2 reduces to the condition that each colour occurs on at least $y - n + 1 = 2$ edges of H_y^{**}. Since each colour occurs on an edge incident with v^{**}, we must now modify the original edge-colouring of H_y^{**} in case that if some colour, say c, is not used on any edge of $H_y^{**} - v^{**}$. In that case, since $|H_y^{**} - v^{**} - v^*| = n + 1$ and $\frac{(n+1)n}{2} - \frac{n+1}{2} \geq 2n - \frac{n+1}{2}$ for any $n \geq 3$, we can find an edge $e^* \notin M$ whose colour occurs on more than one edge of $H_y^{**} - v^{**}$ and which is not adjacent to the edge coloured c incident with v^{**}, and we colour e^* with c. We repeat this as necessary. The argument proceeds as in the case that $e + m = n$. $//$

(**Remarks.** The case that $\bar{G}$ consists of $\frac{1}{2}(n + 1)$ independent edges also follows immediately from Lemma 3.3.)

From the proof of Theorem 6.1, we can come up with an algorithm for obtaining a total colouring of a Type 1 graph G of even order $2n$ having $\Delta(G) = 2n - 1$ using $2n$ colours.

Algorithm. Let $e(\bar{G}) = e$ and $\alpha'(\bar{G}) = m$. By the proof of Hilton's theorem, we may without loss of generality, assume that $e + m = n$ or n is odd and $\bar{G}$ consists of $\frac{1}{2}(n+1)$ independent edges (so that $e + m = n + 1$).

Case 1. $e + m = n$.

1. Label the non-isolated vertices of $\bar{G}$ as $v_1, ..., v_y$. The remaining vertices are labelled as $v_{y+1}, ..., v_{2n}$. Obtain a maximum matching $M = \{v_1 v_{m+1}, v_2 v_{m+2}, ..., v_m v_{2m}\}$ in $\bar{G}$.

2. Construct a graph H^* by adjoining a new vertex v^* to $G + M$ and adding an edge joining v^* to each $v \in \{v_{2m+1}, ..., v_{2n}\}$.

3. Construct a multigraph H^{**} by adjoining a new vertex v^{**} to H^* and adding $2n - d_{H^*}(v)$ edges joining v^{**} to each $v \in V(H^*)$.

4. Let $H_y^{**} = H^{**}[\{v^{**}, v^*, v_1, ..., v_y\}]$. We edge-colour H_y^{**} as follows.

   (a) Colour the edge $v_i v_{m+i}$ with colour c_i $(1 \leq i \leq m)$.

   (b) Colour the $2m$ edges joining v^* and v^{**} with colours $c_1, ..., c_{2m}$.

   (c) Colour the remaining $2n - 2m$ edges on v^{**} with colours $c_{2m+1}, ..., c_{2n}$.

   (d) Colour the remaining edges (if any) of H_y^{**} one after the other in such a way that no adjacent edges receive the same colour.

5. Extend the edge-colouring of H_y^{**} obtained in Step 4 to an edge-colouring of H^{**} using colours $c_1, ..., c_{2n}$ in the following way.

   (a) Construct a bipartite graph B with bipartition $(\{v^{*'}, v_1', ..., v_y'\}, \{c_1', ..., c_{2n}'\})$. A vertex v' is adjacent to a vertex c' if colour c is missing at v in H_y^{**}.

   (b) Find a maximum matching in B with the condition that if a colour c occurs on only $y - n + 1$ edges of H_y^{**}, then c' should be incident with an edge in the matching. (We can make use of the Hungarian Method here.)

   (c) Edge-colour H_{y+1}^{**} as follows: If $c'v'$ is an edge in the maximum matching, then colour $v_{y+1}v$ with c.

   (d) Repeat Steps (a) – (c) unitl we obtain a $2n$-edge-colouring of H^{**}.

6. Remove v^{**} together with its incident edges from H^{**}. We thus have a $2n$-edge colouring φ of H^*.

7. Colour v_i with $\varphi(v^*v_i)$ $(2m + 1 \leq i \leq 2n)$. Colour v_i and v_{m+i} with $\varphi(v_iv_{m+i})$ $(1 \leq i \leq m)$. Delete the edges $v_1v_{m+1}, v_2v_{m+2}, ..., v_mv_{2m}$ and also the vertex v^* together with its incident edges from H^*. We thus have a $2n$-total colouring of G.

Case 2. n is odd and $\bar{G}$ consists of $\frac{1}{2}(n+1)$ independent edges (so that $e + m = n + 1$)

1. Same as Step 1 in Case 1.

2. Same as Step 2 in Case 1.

3. Construct a multigraph H^{**} by adjoining a new vertex v^{**} to H^* and adding $2n - d_{H^*}(v)$ edges joining v^{**} to each $v \in V(H^*)$ except v_1 and v_{m+1}. Add an additional edge joining v_1 and v_{m+1}.

4. Same as Steps 4 (a) − (c) in Case 1. Obtain a 1-factor F such that $F \cap M = \emptyset$. Colour F with colour c_{2m}. Colour the remaining edges such that each colour occurs on at least two edges of H_y^{**}.

5. Same as Step 5 in Case 1.

6. Same as Step 6 in Case 1. Delete the additional edge joining v_1 and v_{m+1} as well.

7. Same as Step 7 in Case 1.

§2. Even order graphs G having $\Delta(G) = |G| - 2$

In this section we shall next classify the graphs G of order $2n$ having $\Delta(G) = 2n - 2$ according to their total chromatic numbers. Surprisingly, a complete classification of such graphs is more specific than that of graphs G of order $2n$ having $\Delta(G) = 2n - 1$ given in Theorem 6.1. This result is due to B. L. Chen and H. L. Fu [92].

In proving their result, Chen and Fu apply the following lemma.

Lemma 6.3 For any spanning star-forest $S = S_{n_1} \cup ... \cup S_{n_k} \neq S_{2n-3} \cup S_1$ or $2S_2$ (if $n = 3$) of K_{2n}, where $n_1 \geq n_2 \geq ... \geq n_k \geq 1$, there exists a (proper) edge-colouring of K_{2n} using $2n - 1$ colours such that all the edges of S receive distinct colours.

The above result is a special case of an unpublished result of L. D. Andersen and E. Mendelsohn (1985). According to Andersen (private communication to the author about three years ago), a full proof of their result (which says that if H is a subgraph of K_{2n}, where $e(H) \leq 2n - 2$, then K_{2n} has a $(2n - 1)$-edge-colouring such that all the edges of H receive distinct colours provided that $H \neq S_{2n-3} \cup S_1$, $K_3 \cup S_1 \cup 0_1$ or $2S_2$) will be no less than 70 pages. At the Third China-USA International Conference on Graph Theory, Combinatorics, Algorithms and Applications held in Beijing, June 1-5, 1993, Mendelsohn told the author that a full proof of the Andersen-Mendelsohn Theorem is still not yet completed. Since Lemma 6.3 is used here, we shall give a direct proof of this result. The proof given here is constructive and thus can be used to provide an algorithm for finding a required edge-colouring. This proof is due to Yap and Liu [-a].

We shall first explain how to use the natural edge-clourings of $K_{n,n}$ and K_n to prove Lemma 6.3. We first split the vertex set of $G(= K_{2n})$ into disjoint union of $X = \{x_1, x_2, ..., x_n\}$ and $Y = \{y_1, y_2, ..., y_n\}$. For convenience we call $G[X]$ and $G[Y]$ the upper K_n and the lower K_n, which are denoted by G_1 and G_2 respectively. We let H be the complete bipartite graph $G(X, Y)$. We shall call the following $(2n - 1)$-edge-colouring of G as a split edge-colouring of G.

Case I. n is even.

We first colour $E(H)$ with n colours $1, 2, ..., n$. The edges of G_1 and G_2 can be coloured independently with $n - 1$ colours $n + 1, ..., 2n - 1$.

Case II. n is odd.

We first colour $E(H) \backslash \{x_1 y_1, x_2 y_2, ..., x_n y_n\}$ with $n - 1$ colours $1, 2, ..., n - 1$ (this can be done using the natural edge-colouring for $K_{n,n}$). We then apply the natural edge-colouring for both G_1 and G_2 using n colours $n, n + 1, ..., 2n - 1$ so that colour $(n - 1) + i$ is missing at vertices x_i and y_i, $i = 1, 2, ..., n$. We then colour the edge $x_i y_i$ with colour $(n - 1) + i$, $i = 1, 2, ..., n$.

In both Case I and Case II we shall be able to produce an edge-colouring of G so that the edges of a copy of $S = S_{n_1} \cup ... \cup S_{n_k}$ sitting inside G receive distinct colours. Lemma 3.3 already proves that Lemma 6.3 is true for $S = nS_1$.

Example 6.1 In the natural edge-colouring φ of K_8, $\varphi(v_1 v_2) = 5$, $\varphi(v_1 v_3) = 2$, $\varphi(v_1 v_5) = 3$, $\varphi(v_8 v_4) = 4$, $\varphi(v_8 v_6) = 6$ and $\varphi(v_8 v_7) = 7$. We observe that $\{v_1 v_2, v_1 v_3, v_1 v_5\} \cup \{v_8 v_4, v_8 v_6, v_8 v_7\}$ forms the edge set of $S = 2S_3$. Hence Lemma

6.3 is true for $S = 2S_3$.

Example 6.2 Applying the split edge-colouring and the natural edge-colourings for $K_{5,5}$ and K_5 we observe that the edges in

$$\{y_1 x_2, y_1 x_5, y_1 y_3, y_1 y_5\} \cup \{x_4 y_2, x_4 y_4, x_4 x_3, x_4 x_1\}$$

receive distinct colours $1, 4, 6, 7, 2, 8, 5$ and 9 (see Fig. 6.1). Hence Lemma 6.3 is true for $S = 2S_4$.

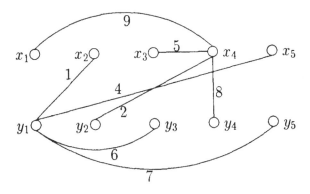

Figure 6.1

To prove Lemma 6.3 we also need to use Lemma 6.4 which is a special case of Theorem 3.5 in Andersen and Hilton [83] (but now stated in the graph terminology).

Lemma 6.4 (Andersen and Hilton [83]) Let G be a balanced complete bipartite graph of order $2n$. Suppose $S = S_{m_1} \cup ... \cup S_{m_k}$, where $m_1 + ... + m_k = n$, is a subgraph of G. Then any edge-colouring of S with n distinct colours can be completed to a (proper) edge-colouring of G using the same set of n colours provided that one of the following holds

(i) $k = 1$ or $k \geq 3$

(ii) when $k = 2$, $\min\{m_1, m_2\} \geq 2$ and the two centres of S_{m_1} and S_{m_2} are in the same partite set.

Proof of Lemma 6.3

Let $G = K_{2n}$. Since S is a spanning subgraph of G,

$$n_1 + n_2 + \dots + n_k = 2n - k.$$

We first make an important note here: In the proof we need only to consider an isomorphic copy of S sitting in G.

This Lemma is clearly true if $k = 1$. Hence we assume that $k \geq 2$. Since we had already settled the case $S = nS_1$, we shall from now on assume that $S \neq nS_1$.

We shall use induction and Lemma 6.4 to find a required split edge-colouring for G. Let the centre of S_{n_i} be z_i, and let $w_i z_i$ be an edge of S_{n_i}, $i = 1, 2, \dots, k$.

We first settle the case that $k = 2$. We place z_i at y_i, $i = 1, 2$.

Suppose $n_2 = 2$. Then $n_1 + n_2 = 2n - 2$ implies that $n_1 = 2n - 4$. Since $S_{n_1} \cup S_{n_2} \neq S_{2n-3} \cup S_1$ or $2S_2$ (if $n = 3$), we have $n \geq 4$ and thus $n_1 = 2n - 4 \geq 4$. We now place all the two edges of S_{n_2} in H, $n-2$ edges of S_{n_1} in H, and the remaining $n_1 - (n-2) = (2n-4) - (n-2) = n - 2 \geq 2$ edges of S_{n_1} in G_2. By Lemma 6.4, there exists an n-edge-colouring φ_1 of H using colours $1, 2, \dots, n$ such that all the n edges of $E(S_{n_1} \cup S_{n_2}) \cap E(H)$ receive distinct colours.

Suppose n is even. Since $E(S_{n_1} \cup S_{n_2}) \cap E(G_2)$ induces a star of size $n - 2$ in G_2, there exists an $(n-1)$-edge-colouring φ_2 of G_1 and G_2 using $n - 1$ colours $n+1, \dots, 2n-1$ such that all the edges in $E(S_{n_1} \cup S_{n_2}) \cap E(G_2)$ receive distinct colours in $n + 1, \dots, 2n - 1$. Hence a required $(2n - 1)$-edge-colouring of G exists.

Suppose n is odd. We do almost entirely the same process as the case when n is even except that now we assume that in φ_1 the edges $x_1 y_1, x_2 y_2, \dots, x_n y_n$ are coloured with colour n, and in particular, $w_1 z_1 = y_1 x_1$ is coloured with colour n. We identify $x_1, x_2, \dots, x_n$ as a single vertex x^*. We let φ_2 be an edge-colouring of the complete graph having vertex set $\{y_1, y_2, \dots, y_n, y_{n+1} = x^*\}$ using n colours $n, n+1, \dots, 2n-1$ such that

$$\varphi_2(x^* y_i) = (n-1) + i, \quad i = 1, 2, \dots, n.$$

We also let φ_3 be an edge-colouring of G_1 using n colours $n, n+1, \dots, 2n-1$ such that colour $(n-1) + i$ is missing at vertex x_i, $i = 1, 2, \dots, n$. We recolour the edge $x_i y_i$

with colour $(n-1)+i$ for all $i = 1, 2, ..., n$. Then we obtain a requird edge-colouring for G.

Suppose $n_2 = 3$. If $n = 4$, then $S = 2S_3$ which had been settled in Example 6.1. Hence we assume that $n \geq 5$. Now again we place all the three edges of S_{n_2} in H, $n - 3 \geq 2$ edges of S_{n_1} in H and all the remaining $n - 2$ edges of S_{n_1} in G_2. The rest of the proof is exactly the same as for the case $n_2 = 2$.

Suppose $n_2 = 4$. We place two edges of S_{n_2} in H and the other two edges of S_{n_2} in G_2. We next place $n - 2 (\geq 3)$ edges of S_{n_1} in H and the remaining $n_1 - (n - 2)$ edges of S_{n_1} in G_2. This case can be settled exactly the same as the case $n_2 = 2$ except when $n_1 - (n - 2) = 1, 2$, i.e. $S = 2S_4$ and $S = S_6 \cup S_4$. The case that $S = 2S_4$ had been settled in Example 6.2. The case that $S = S_6 \cup S_4$ will be settled in Example 6.3 below.

For $n_2 \geq 5$, we can place either two or three edges of S_{n_2} in H and the rest of the proof is exactly the same as for the case $n_2 = 2$ except that $S = S_5 \cup S_5$ which we shall settle in Example 6.4 below.

From now on we assume that $k \geq 3$. Since $S \neq nS_1$, we have $n \geq k + 1 \geq 4$. We first place z_1 at y_1.

Suppose $n_1 \geq n - (k - 1)$. We now place z_i at y_i, w_i at x_i for all $i = 2, 3, ..., k$. We next place $n - (k - 1)$ edges of S_{n_1} (including $w_1 z_1$) in H. Then we place all the remaining edges of $S_{n_1}, S_{n_2}, ...,$ and S_{n_k} in G_2. Observe that

$$S' = S_{n-(k-1)} \cup S_1 \cup ... \cup S_1 \ (k - 1 \text{ copies } S_1)$$

is a vertex-disjoint union of stars in H, where $e(S') = n$. By Lemma 6.4, there exists an n-edge-colouring φ_1 of H using colours $1, 2, ..., n$ such that all the n edges of S' receive distinct colours. The remaining edges of S form

$$S'' = S_{n_1-(n-(k-1))} \cup S_{n_2-1} \cup ... \cup S_{n_k-1}$$

in G_2, where $e(S'') = n - k$.

If $n = 4$, the result is clearly true because $e(S'') = 1$. Suppose $n \geq 6$ is even. Then the result follows by applying Lemma 1 and the induction hypothesis on G_2. (Note that for $k \geq 4$, we can apply the induction hypothesis without any difficulty.

For instance, if $S'' = S_3 \cup S_1 \cup 2S_0$, then $S'' \subseteq S_3 \cup 2S_1$ and since the result is true for $S_3 \cup 2S_1$ it must be true for $S'' = S_3 \cup S_1 \cup 2S_0$ also. However, for $k = 3$ we may encouter difficulties in applying the induction hypothesis if $S'' = S_2 \cup S_1 \cup S_0$. But in this case $S = S_6 \cup S_2 \cup S_1$ and so we may place all the two edges of S_{n_2} in H, i.e. $S' = S_3 \cup S_2 \cup S_1$ and so $S'' = S_3 \cup 2S_0 \subseteq S_5$ and thus we can apply the induction hypothesis also.)

Suppose $n \geq 5$ is odd. We first identify the vertices $x_1, x_2, ..., x_n$ into a single vertex x^*. Then $\{y_1, y_2, ..., y_n, y_{n+1} = x^*\}$ forms a complete graph K_{n+1} which can be edge-coloured with n colours $n, n+1, ..., 2n-1$. Now $|S'' \cup \{x^* y_1\}| = n - k \leq n - 3$ and thus we can also apply the induction hypothesis (and Lemma 6.4) to obtain a required split edge-colouring for G except when $S'' = 2S_1 \cup S_0$. However, in this case $S = S_4 \cup S_2 \cup S_1$ and we can place all the two edges of S_{n_2} in H so that $S'' = S_2 \cup 2S_0 \subseteq S_4$ and thus we can apply the induction hypothesis on the complete graph induced by $\{y_1, ..., y_5, y_6 = x^*\}$ also. We next let $\varphi_1(y_1 x_1) = n$ and let $\{x'_1, ..., x'_n\}$ be a permutation of $\{x_1, x_2, ..., x_n\}$ such that $x'_1 = x_1$ and $\{x'_1 y_1, ..., x'_n y_n\}$ is the colour class of φ_1 which receive colour n in φ_1. As before, we let $(n+1) + j$ be the colour absent at vertices y_j and x'_j in φ_2 and φ_3, we can recolour the edge $y_j x'_j$ by colour $(n-1) + j$ to obtained a required $(2n-1)$-edge-colouring of G.

Suppose $n_1 < n - (k-1)$. In this case, we place all the n_1 edges of S_{n_1} in H. Now if $n_2 \geq n - (n_1 + (k-2))$, we place $n - (n_1 + (k-2))$ edges of S_{n_2} in H and all the remaining edges of $S_{n_2}, S_{n_3}, ..., S_{n_k}$ in G_2. Using a similar argument we can obtain a required edge-colouring for G. On the other hand, if $n_2 < n - (n_1 + (k-2))$, then we also place all the edges of S_{n_2} in H and we continue the same process for S_{n_3} and so on and eventually we will obtain a required edge-colouring of G by applying Lemma 6.4 and the induction hypothesis on $E(S) \cap E(G_2)$.   //

In the proof of Lemma 6.3, we used the following two special cases $S = S_6 \cup S_4$ and $S = 2S_5$. We now settled these two special cases.

Example 6.3 K_{12} has an edge-colouring using colours $1, 2, ..., 11$ such that all the edges of a spanning star-forest $S = S_6 \cup S_4$ receive distinct colours.

We first split $V(K_{12}) = \{x_1, ..., x_6\} \cup \{y_1, ..., y_6\} = X \cup Y$. Let φ_1 be the natural edge-colouring of $G(X, Y)$ using colours $1, 2, ..., 6$. Let φ_2 be the natural edge-colouring of G_2 using colours $7, 8, ..., 11$. Let φ_3 be the natural edge-colouring of G_1 using colours $7, 8, ..., 11$ in which vertices x_2 and x_5 are interchanged and vertices x_3 and x_4 are also interchanged.

Observe that $\{y_1x_2, y_1x_3, y_1x_4, y_1y_4, y_1y_5, y_1y_6\}$ and $\{x_1y_2, x_1y_3, x_1x_5, x_1x_6\}$ form the edge sets of S_6 and S_4. The edges of $S_6 \cup S_4$ are thus coloured with colours $1, 2, 3, 11, 9, 7, 5, 4, 10$ and 8 (see Fig. 6.2). Hence Lemma 6.3 holds for $S_6 \cup S_4$.

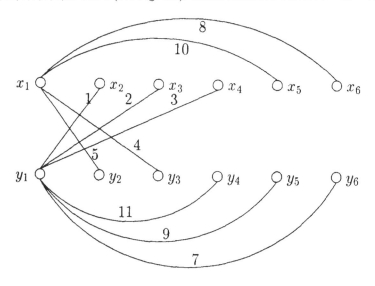

Figure 6.2

Example 6.4 K_{12} has an edge-colouring using colours $1, 2, .., 11$ such that all the edges of a spanning star-forest $S = 2S_5$ receive distinct colours.

We first split $V(K_{12}) = \{x_1, ..., x_6\} \cup \{y_1, ..., y_6\} = X \cup Y$. By Lemma 6.4, there exists an edge-colouring of $G(X, Y)$ using colours $1, 2, ..., 6$ such that the edges $y_1x_1, y_1x_2, y_1x_3, y_5x_4, y_5x_5, y_5y_6$ receive distinct colours $1, 2, ..., 6$ respectively.

The natural colouring φ_2 of G_2 using colours $7, 8, 9, 10$ and 11 is such that $\varphi_2(y_1y_2) = 10$, $\varphi_2(y_1y_3) = \varphi_2(y_5y_4) = 8$ and $y_2(y_5y_6) = 11$. Without loss of generality we assume that $\varphi_1(y_5x_2) = 1$ and we let $x' \in \{x_3, x_4, x_5, x_6\}$ be such that $\varphi_1(y_4x') = 1$. We also observe that $\varphi_2(y_1y_6) = 7$ (see Figure 6.3). Let $x'' \in \{x_3, x_4, x_5, x_6\} \setminus \{x'\}$ be such that $\varphi_1(y_6x'') = 1$. Finally, we let φ_3 be an edge colouring of $G[X]$ using colours $7, 8, 9, 10$ and 11 satisfying $\varphi_3(x_2x') = 8$ and

$\varphi_3(x_1 x'') = 7$. Interchanging the colours 8 and 1 in the cycle $y_4 y_5 x_2 x' y_4$ and the colours 7 and 1 in the cycle $y_1 y_6 x'' x_1 y_1$, we obtain an 11-edge-colouring of K_{12} such that the edges in $\{y_1 x_1, y_1 x_2, y_1 x_3, y_1 y_2, y_1 y_3\} \cup \{y_5 x_4, y_5 x_5, y_5 x_6, y_5 y_4, y_5 y_6\}$ receive distinct colours.

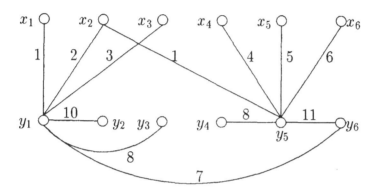

Figure 6.3

(**Remarks.** We can deduce a parallel result of Lemma 6.3 for the existence of a $(2n - 1)$-edge-colouring of K_{2n-1} so that all the edges of spanning star forests $S = S_{n_1} \cup ... \cup S_{n_k}$ receive distinct colours where $n_1 \geq n_2 \geq ... \geq n_k$ with possibly $n_{k-1} > n_k = 0$.)

We next introduce some notation and prove a few more lemmas. The union of k vertex-disjoint stars $S_{n_1}, ..., S_{n_k}$ (S_{n_i} is a star of size n_i) is denoted by $S(n_1, ..., n_k)$ and the complement of $S(n_1, ..., n_k)$ is denoted by $G(n_1, ..., n_k)$.

Lemma 6.5 Let G be a graph of order $2n$ having $\Delta(G) = 2n - 2$. Then $G \subseteq G(n_1, ..., n_k)$ for some integers $n_i \geq 1$, $i = 1, 2, ..., k$ such that

$$k + (n_1 + n_2 + ... + n_k) = 2n.$$

Proof. Since $\Delta(G) = 2n - 2$, $\bar{G}$ contains a spanning forest F. From F, by deleting edges whose end-vertices are both of degree at least two, we obtain a spanning star forest $S(n_1, ..., n_k)$, where each $n_i \geq 1$ and $k + (n_1 + n_2 + ... + n_k) = 2n$. Hence $G \subseteq G(n_1, ..., n_k)$. //

Lemma 6.6 For any $n \geq 2$, the graph $G(2n - 3, 1)$ is Type 2.

Proof. Let $G = G(2n - 3, 1)$, $\bar{G} = S_{2n-3} \cup S_1$ and $V(S_1) = \{u, v\}$. Suppose $\chi_T(G) = \Delta(G) + 1 = 2n - 1$. Let π be a $(2n - 1)$-total-colouring of G using colours $c_1, c_2, ..., c_{2n-1}$. Let w be the centre of S_{2n-3}.

Case 1. $\pi(u) = \pi(v) = c_j$, say.

Let $\pi(w) = c_1$, $\pi(wu) = c_2$, and $\pi(wv) = c_3$. Clearly $j \geq 4$. Let r_i be the number of vertices y for which colour c_i occurs (i.e. either $\pi(y) = c_i$ or $\pi(e) = c_i$ for some edge e incident with y). Then $r_1, r_2, r_3 \leq 2n$, $r_j \leq 2n - 2$ and $r_i \leq 2n - 1$ for any $i \neq 1, 2, 3, j$. Hence

$$4n^2 - 4n + 4 = v(G) + 2e(G) = r_1 + r_2 + ... + r_{2n-1}$$
$$\leq 3(2n) + (2n - 2) + (2n - 5)(2n - 1) = 4n^2 - 4n + 3,$$

which is false.

Case 2. $\pi(u) \neq \pi(v)$.

In this case G has a vertex $x \neq u, v, w$ such that $\pi(x) = \pi(w) = c_1$, say. Similar to Case 1, we have

$$4n^2 - 4n + 4 = v(G) + 2e(G) = r_1 + r_2 + ... + r_{2n-1}$$
$$\leq 2n + (2n - 2)(2n - 1) = 4n^2 - 4n + 2,$$

which is false.

Finally, since $2n - 1 \leq \chi_T(G) \leq 2n$ (see Exercise 5(1)), we now have $\chi_T(G) = 2n$. //

Lemma 6.7 If $k + (n_1 + n_2 + ... + n_k) = 2n$ and $G(n_1, ..., n_k) \not\cong G(2n - 3, 1)$, then $G(n_1, ..., n_k)$ is Type 1.

Proof. By Theorem 3.1, $G(n_1) = K_{2n-1} \cup O_1$ is Type 1. Suppose $k \geq 2$ and $n \geq 4$. Then by Lemma 6.3, K_{2n} has a $(2n - 1)$-edge-colouring φ such that all the $2n - k$ edges $e_1, ..., e_{2n-k}$ in $S(n_1, ..., n_k) = \overline{G(n_1, ..., n_k)}$ receive distinct colours. We now modify φ to a total-colouring of $G(n_1, ..., n_k)$ using the same set of $2n - 1$ colours as follows: The colouring φ remains an edge-colouring of $G(n_1, ..., n_k)$. For each star S_{n_j} in $S(n_1, ..., n_k)$ with $V(S_{n_j}) = \{v_j^{(0)}, v_j^{(1)}, ..., v_j^{(n_j)}\}$ we put $\varphi(v_j^{(0)}) = \varphi(v_j^{(1)}) = \varphi(v_j^{(0)} v_j^{(1)})$ and $\varphi(v_j^{(i)}) = \varphi(v_j^{(0)} v_j^{(i)})$. It is easy to check that φ is a total-colouring of $G(n_1, ..., n_k)$. The remaining case $n \leq 3$ can be checked easily. $\quad //$

Lemma 6.8 If H is a proper induced subgraph of $G = G(2n - 3, 1)$ with $\Delta(H) = 2n - 2$, then H is Type 1.

Proof. It suffices to show that for any edge e of G, $G - e$ is Type 1. Let $V(G) = \{v_0, v_1, ..., v_{2n-1}\}$. Suppose v_{2n-2} is not adjacent to v_{2n-1} and v_0 is not adjacent to $v_1, v_2, ..., v_{2n-3}$. By symmetry, we need only to consider three cases.

In case $e = v_1 v_{2n-1}$ and $n \geq 4$, $G - e + v_0 v_1 \simeq G(2n - 4, 2)$, which by Lemma 6.7, is Type 1. In case $e = v_1 v_2$, $G - e + v_0 v_1 + v_0 v_2 \simeq G(2n - 5, 1, 1)$, which by Lemma 6.7, is Type 1. Hence, by Lemma 2.1, $G - e$ is Type 1.

In case $e = v_0 v_{2n-1}$, a $(2n - 1)$-total-colouring π of $G - e$ can be obtained by modifying a $(2n - 1)$-total-colouring φ of K_{2n-1}. Let $p = 2n - 1$ and let the vertex set of K_p be $\{1, 2, ..., p\}$. The p colour classes of φ are

$$C(j) = \{\{j - i, j + i\} | i = 1, 2, ..., p - 1\} \cup \{j\}, \quad j = 1, 2, ..., p,$$

where $j - i$ and $j + i$ are calculated modulo p. The colour classes of π are $C(1) \cup \{2n\}$, $C(j)$ for $j = 2, 3, ..., p - 1$ and $C(p) \setminus \{n - 1, n\} \cup \{n, 2n\}$.

Finally for the case $e = v_1 v_{2n-1}$ and $n = 3$, it is not difficult to establish a $(2n - 1)$-total-colouring for G. $\quad //$

Combining Lemma 6.5, Lemma 6.6, Lemma 6.6 and Lemma 6.8, we have

Theorem 6.9 (Chen and Fu [92]) Let G be a graph of order $2n$ having $\Delta(G) = 2n - 2$. Then G is Type 2 if and only if $\bar{G} = S_{2n-3} \cup S_1$.

§3. Odd order graphs G having $\Delta(G) = |G| - 2$

In this section we shall classify the graphs G is of odd order $2n + 1$ having $\Delta(G) = 2n - 1$ according to their total chromatic numbers. This case is much more difficult than the parallel case that G is of even order $2n$ studied by Chen and Fu.

To give a complete classification of graphs G of order $2n + 1$ (according to their total chromatic members) having $\Delta(G) = 2n - 1$, we first have to prove many preliminary theorems and lemmas.

Theorem 6.10 Let G be a graph of order $2n + 1$ having $\Delta(G) = 2n - 1$. If G is Type 1, then for any $w \in V(G)$,

$$(2) \qquad e(\overline{G - w}) + \alpha'(\overline{G - w}) \geq n.$$

Proof. Let $\Delta = \Delta(G)$. Suppose $\Delta(G - w) = \Delta$. Then

$$\Delta + 1 = \Delta(G - w) + 1 \leq \chi_T(G - w) \leq \chi_T(G) = \Delta + 1$$

from which it follows that $\chi_T(G - w) = \Delta(G - w) + 1$. Hence, by Theorem 6.1, (2) holds.

Suppose $\Delta(G - w) = \Delta - 1$. Since $|G - w| = 2n$ and $\Delta(G - w) = 2n - 2$, each vertex in $\overline{G - w}$ is of degree at least one. Hence $e(\overline{G - w}) \geq \frac{1}{2} |\overline{G - w}| = n$, and (2) holds also. //

Theorem 6.11 Let G be a graph of order $2n + 1$ having $\Delta(G) = 2n - 1$. If G has a vertex w such that

$$(3) \qquad e(\overline{G - w}) + \alpha'(\overline{G - w}) = n,$$

then G is Type 1.

Proof. Let $\Delta = \Delta(G)$. If $\Delta(G - w) = \Delta - 1$, then $(\Delta - 1) + 2 = 2n$ and by Theorem 5.5, $G - w$ has a $2n$-total-colouring φ. If $\Delta(G - w) = \Delta$, then by Theorem 6.1, $G - w$ also has a $2n$-total-colouring φ. Now

$$2e(G - w) + |G - w| = 2[n(2n - 1) - e(\overline{G - w})] + 2n$$

$$= 4n^2 - 2e(\overline{G - w}) = 4n^2 - 2(n - \alpha'(\overline{G - w})) = 4n^2 - 2n + 2p,$$

where $p = \alpha'(\overline{G - w})$.

Clearly there are $q \geq 2n - 2p$ colours (say colours $1, 2, ..., q$) each of which is used to colour exactly one vertex of $G - w$. Since the total number of colour-vertex pairs is $4n^2 - 2n + 2p$ and is bounded by $q(2n - 1) + (2n - q)(2n)$, we have

$$4n^2 - 2n + 2p \leq q(2n - 1) + (2n - q)(2n) = 4n^2 - q$$

from which it follows that $q \leq 2n - 2p$. Hence $q = 2n - 2p$, and from the above inequality we know that G has a $2n$-total-colouring φ such that:

(i) the first $2n - 2p$ colours $1, 2, ..., 2n - 2p$ were used to colour $2n - 2p$ vertices, one colour for one vertex;

(ii) the next p colours $2n - 2p + 1, ..., 2n - p$ were used to colour $p = \alpha'(\overline{G - w})$ pairs of vertices of independent edges in $\overline{G - w}$, and each such colour is present at every vertex of $G - w$;

(iii) the last p colours $2n - p + 1, ..., 2n$ were used to colour only edges of $G - w$ so that each such colour is present at every vertex of $G - w$;

(iv) the colours missing at the minor vertices of $G - w$ are the colours from $\{1, 2, ..., 2n - 2p\}$, each such colour is missing at only one vertex.

Hence we can extend φ to a $2n$-total-colouring of G using the same set of $2n$ colours $\{1, 2, ..., 2n\}$ by putting $\varphi(wx) = \alpha$ if α is absent at vertex x, and $\varphi(w) = \beta$ for any $\beta \in \{2n - p + 1, ..., 2n\}$.   //

The proof of Theorem 6.11 gives us a lot of information on the $2n$-total-colouring φ of $G - w$. Our proof of the next main result (Theorem 6.23) of this section depends heavily on this information. Here we give an example (Example 6.5) to illustrate the total-colouring φ of G. We recall that the graph $G(n_k, n_{k-1}, ..., n_1)$ is a graph of order $2n + 1 = k + (n_1 + n_2 + ... + n_k)$ if $\bar{G} = S_{n_k} \cup S_{n_{k-1}} \cup ... \cup S_{n_1}$, where S_{n_i} is a star of size n_i, and $n_k \geq n_{k-1} \geq ... \geq n_1$.

Example 6.5 Let $G = G(5, 2, 1)$. A 10-total-colouring of G is partially depicted in Fig.6.4. The dotted lines in Fig.6.4 are non-edges of G. The colours $\alpha_1, \alpha_2, ..., \alpha_5 \in \{1, 2, ..., 6\}$ are the colours missing at the respective vertices of $G - w$.

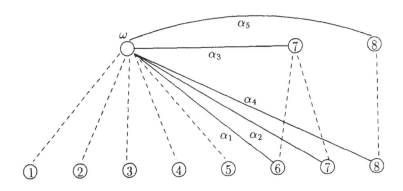

Figure 6.4

We shall require the following two results to prove Theorem 6.14.

Theorem 6.12 (V. Chvátal (1972); C. Berge (1973)). Let G be a graph of order p having minimum degree δ. If $\delta \geq \frac{1}{2}(p + r)$, where r is a positive integer, then any r edges which form disjoint paths in G are contained in a Hamilton cycle of G.

(A proof of this theorem can be found in several graph theory textbooks.)

Theorem 6.13 (Chetwynd and Hilton [84]) Suppose G is a graph of order $2n+2 \geq 6$ and $\Delta(G) = 2n - 1$. Then G is Class 2 if and only if

$$e(G - w) > \binom{2n + 1}{2} - 2n,$$

where w is a vertex of minimum degree.

Theorem 6.14 Suppose G is a connected graph of odd order $2n + 1 \geq 7$, $\Delta(G) = 2n - 1$, G is maximal, and

$$(4) \quad e(\bar{G}) - \Delta(\bar{G}) \geq \begin{cases} n - 1 & \text{if } G \text{ has only one vertex of minimum degree,} \\ n & \text{if } G \text{ has at least two vertices of minimum degree.} \end{cases}$$

Then $\chi_T(G) = \Delta(G) + 1$.

Proof. We first note that since $|G| = 2n + 1$ and $\Delta(G) = 2n - 1$ are both odd, G cannot be a regular graph. It is important to note that since $\Delta(G) = 2n - 1$, each major vertex of G is of degree 1 in $\bar{G}$.

Let w be a vertex of degree $\delta = \delta(G)$ in G. From condition (4) it follows that $e(\overline{G - w}) \geq n - 1 \geq 2$. Since G is maximal, the end-vertices of each edge in $\overline{G - w}$ are adjacent to w in G.

If $\alpha'(\overline{G - w}) \geq 2$, then $\overline{G - w}$ has three distinct vertices z, x and y such that z is of minimum degree $\delta(G - w)$ in $G - w$, $xy \in E(\overline{G - w})$, and y is a major vertex of G.

We next prove that if $\alpha'(\overline{G - w}) \geq 2$, then $d(z) \geq n + 2$. Suppose otherwise. Then $d(w) \leq d(z) \leq n + 1$ and thus $d_{\bar{G}}(w) \geq n - 1$. Since each $u \in N_{\bar{G}}(w)$ is a major vertex in G, we have $uz \in E(G)$. Suppose $zx \notin E(G)$. Then x cannot be a major vertex in G, and since G is maximal, z must be a major vertex in G and thus $d(z) = 2n - 1 \geq n + 2$, a contradiction. Hence $zx \in E(G)$. Now $w, x, y \in N(z)$ implies that $d(z) \geq d_{\bar{G}}(w) + 3 \geq n + 2$, another contradiction. Consequently, $d(z) \geq n + 2$ always.

Let $G' = G - w + xy$. Since

$$d_{G'}(z) = \delta(G') \geq n + 1 = \frac{1}{2}(|G'| + 1),$$

by Theorem 6.12, G' has a Hamilton cycle C containing the edge xy from which it follows that $G - w$ has a matching M such that x and y are the only two M-unsaturated vertices.

Suppose $\alpha'(\overline{G - w}) = 1$. Then the nontrivial connected component of $\overline{G - w}$ is a star. Let the centre of this star be x and let y be a vertex not adjacent to x in G. In this case $G - \{w, x, y\} \simeq K_{2n-2}$ and thus has a perfect matching M.

Next, let G^* be the graph obtained from $G - M$ by adjoining a new vertex v^* and adding an edge joining v^* to each vertex in $V(G - M) \setminus \{x, y\}$. Then $v(G^*) = 2n + 2$ and $\Delta(G^*) = 2n - 1$. Now if

$$(5) \qquad e(G^* - w') \leq \binom{2n + 1}{2} - 2n = n(2n - 1),$$

where w' is a vertex of minimum degree in G^*, then, by Theorem 6.13, G^* is Class 1 and hence G is Type 1.

Finally we show that condition (5) holds. Suppose w is the only vertex of minimum degree in G. Then w is a vertex of minimum degree in G^* and condition (5) (when w' is replaced by w) is equivalent to

$$e(G) - |M| + (2n - 1) - (\delta(G) + 1) \leq n(2n - 1)$$

$$\Longleftrightarrow [(2n + 1)n - e(\bar{G})] - (n - 1) + (2n - 1) - [(2n - \Delta(\bar{G}) + 1] \leq n(2n - 1)$$

$$\Longleftrightarrow e(\bar{G}) - \Delta(\bar{G}) \geq n - 1.$$

Similarly, if G has at least two vertices of minimum degree $\delta(G)$, then $\delta(G^*) = \delta(G)$ and condition (5) holds if and only if $e(\bar{G}) - \Delta(\bar{G}) \geq n$. $//$

Example 6.6 Let G be a graph on 7 vertices. Suppose one of the following holds.

(i) $\bar{G} = P_2 \cup P_5$ or $P_3 \cup P_4$ (ii) $\bar{G} = 2P_2 \cup P_3$ (iii) $\bar{G} = S_2 \cup S_3$.

Then by Theorem 6.14, G is Type 1.

Since $e(\bar{G}) - \Delta(\bar{G}) = e(\bar{G}) - (2n - \delta(G)) = e(\bar{G}) + \delta(G) - 2n$, we know that condition (5) is equivalent to

$$(5') \qquad e(\bar{G}) + \delta(G) \geq \begin{cases} 3n - 1 & \text{if } G \text{ has only one vertex of minimum degree,} \\ 3n & \text{otherwise.} \end{cases}$$

We can show that $G = G(6, 3, 1, 1)$ is Type 1 but $e(\bar{G}) + \delta(G) = 19 \ngeq 3 \times 7 - 1$. Hence condition $(5')$ is a sufficient condition for G to be Type 1 but it is not a necessary condition.

We shall require Theorem 6.14 and the following lemma (together with its corollary) to prove Theorem 6.17.

Lemma 6.15 Let G be a graph of order $2n + 1$ having $\Delta(G) = 2n - 1$. Suppose w is a vertex of minimum degree in G. If all the vertices in $N_G(w)$ are minor vertices in G, then

$$e(\bar{G}) + \delta(G) \geq 3n$$

and thus G is Type 1.

Proof. Since each vertex v in $\overline{G - w}$ is of degree at least one, we have

$$e(\overline{G - w}) \geq \frac{1}{2} |\overline{G - w}| = n.$$

Let r be the number of vertices of G which are not adjacent to w in G. Then

$$e(\bar{G}) = e(\overline{G - w}) + r \geq n + r.$$

Hence $e(\bar{G}) + \delta(G) \geq (n + r) + (2n - r) = 3n$ and, by Theorem 6.14, G is Type 1. //

Corollary 6.16 Let G be a graph of order $2n + 1$ having $\Delta(G) = 2n - 1$. Suppose w is a vertex of minimum degree in G. Then $\alpha'(\overline{G - w}) = \alpha'(\bar{G}) - 1$. Otherwise

$$\mathbf{e}(\bar{G}) + \delta(G) \geq 3n.$$

Proof. If G has a major vertex which is not adjacent to w, then clearly $\alpha'(\overline{G - w}) = \alpha'(\bar{G}) - 1$. //

Theorem 6.17 To prove that every graph G of order $2n + 1$ having $\Delta(G) = 2n - 1$ is Type 1 if and only if (3) holds, it suffices to prove it only for graphs $G(n_k, n_{k-1}, ..., n_1)$.

Proof. Let w be a vertex of minimum degree in G. We are given that

$$e(\overline{G - w}) + \alpha'(\overline{G - w}) = n + s, \quad s \geq 0.$$

By Theorem 6.11, we can assume that $s \geq 1$. By Theorem 6.1 we can assume that G is connected. By Lemma 6.15 we can assume that at least one vertex in $N_G(w)$ is a major vertex of G. Let $p = \alpha'(\overline{G - w})$ and let $M = \{x_1 y_1, ..., x_p y_p\}$ be a set of independent edges in $\overline{G - w}$. By adding an edge $e \notin M$ joining two non-adjacent minor vertices of G in $G - w$, we reduce the value of $e(\overline{G - w}) + \alpha'(\overline{G - w})$ by 1. We add such edges one-by-one until we end up with a supergraph G' of G such that $e(\overline{G' - w}) + \alpha'(\overline{G' - w}) = n$ or $e(\overline{G' - w}) + \alpha'(\overline{G' - w}) > n$ but there are no more such edges. In the former case, G' is Type 1 and thus G is Type 1. Hence we assume that we have the later case.

Next, if $u \in N_G(w)$ has $d_{G'}(u) < \Delta$, we add an edge joining u and w. Let G^* be the graph obtained from G' by adding all such edges. Since at least one vertex in $N_{\bar{G}}(w)$ is a major vertex of G, we have $d_{G^*}(w) \leq \Delta$ and $\bar{G}^* = S_{n_k} \cup ... \cup S_{n_1}$.

Now if w is of minimum degree in G^*, then the proof of Theorem 6.17 is complete. Suppose otherwise. Let w^* be a vertex of minimum degree in G^*.

Suppose $e(\overline{G^* - w^*}) + \alpha'(\overline{G^* - w^*}) < n$. Then $d_{G^*}(w) \geq d_{G^*}(w^*) + 1 \geq (2n - n_k) + 1$ (note that n_k is the degree of w^* in G^*) and so at some stage when we were adding in edges uw where $u \in N_{\bar{G}}(w)$ and $d_{G'}(u) < \Delta$, we would have obtained a supergraph G'' of G' such that $e(\overline{G'' - w^*}) + \alpha'(\overline{G'' - w^*}) = n$. Hence by Theorem 6.11, G'' is Type 1 and so G is Type 1 (because G'' is a supergraph of G). Hence we assume that $e(\overline{G^* - w^*}) + \alpha'(\overline{G^* - w^*}) \geq n$ and the proof of Theorem 6.17 is complete. $//$

We observe that for any graph $G = G(n_k, n_{k-1}, ..., n_1)$ of order $2n + 1$, in which w is of minimum degree

$$e(\overline{G - w}) + \alpha'(\overline{G - w}) = n + s, \; s \geq 1$$

if and only if $(2n + 1 - k - n_k) + (k - 1) = n + s$, i.e. if and only if

$$(6) \hspace{4cm} n_k = n - s.$$

Theorem 6.18 Let $G = G(n_k, n_{k-1}, ..., n_1)$ be of order $2n + 1$, where $n_k = n - 1$ and $n_{k-1} = n_{k-2} = ... = n_1 = 1$. Then G is Type 1.

Proof. Here we have $n_k = n - 1$ and $k - 1 = \frac{1}{2}(n + 1)$, n odd.

Let $p = k - 1$. Let w be the centre of S_{n_k} in $\bar{G} = S_{n_k} \cup S_1 \cup ... \cup S_1$ and let $x_1 y_1, ..., x_p y_p$ be the p edges in $\overline{G - w}$. Let $\bar{G}^*$ be the graph obtained from $\bar{G}$ by deleting $x_1 y_1$ and adding two edges $w x_1$ and $x_2 y_1$. Then

$$e(\overline{G^* - w}) + \alpha'(\overline{G^* - w}) = n.$$

By Theorem 6.11, G^* has a $2n$-total-colouring φ which is partially depicted in Fig. 6.5.

Using the total-colouring given in the proof of Theorem 6.11, we let the $n + 1$ colours $1, 2, ..., n+1$ missing at the vertices $x_2, x_3, ..., x_p, y_1, y_2, ..., y_p$ be $\alpha_1, \alpha_2, ..., \alpha_{n+1}$ respectively. (Note that there are two colours missing at vertex x_2, namely α_1 and α_2.) Let $w_1, ..., w_{n-1}$ be the end-vertices of S_{n_k} and let $w_n = x_1$. If $\varphi(y_1 w_i) = \alpha_1$ for some $i \leq n$, say $\varphi(y_1 w_n) = \alpha_1$ (there is no loss of generality to select $y_1 w_n$ to be the edge coloured α_1 because we are dealing with isomorphic copies of G), then we can easily modify φ to a $2n$-total-colouring of G by setting $\varphi(x_2 y_1) = \alpha_1$, deleting the edge $x_1 y_1$, and extending φ using the technique given in the proof of Theorem 6.11.

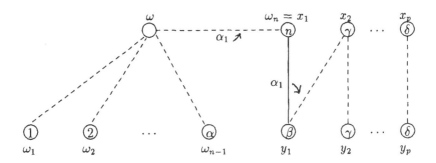

$$(\alpha = n - 1, \beta = n + 1, \gamma = n + 2, \delta = n + p)$$

Figure 6.5

For convenience in the later part of the proof, we let $\{z_1, ..., z_{2p-1}\} = \{x_2, ..., x_p, y_1, ..., y_p\}$. We next assume that $\varphi(y_1 z_i) = \alpha_1$ for some i. Again we shall modify φ to a $2n$-total-colouring φ (abusing the use of notation) of $G^* + x_2 y_1 - y_1 z_i$. Using a similar argument, we can show that $\varphi(y_1 z_j) = \alpha_\ell$ for some z_j, where α_ℓ is the colour missing at z_i. By performing the same "fan-recolouring process", where the fan is pivoted at y_1, we have shown that all the colours $\alpha_1, \alpha_2, ..., \alpha_{n+1}$, except the colour missing at y_1, are colours in

$$S = \{\varphi(y_1 x_3), \varphi(y_1 x_4), ..., \varphi(y_1 x_p), \varphi(y_1 y_2), ..., \varphi(y_1 y_p)\}.$$

However, $|S| = (p - 2) + (p - 1) = 2p - 3 < 2p - 2 = n - 1$. Hence at some stage in the "fan-recolouring process", we must have $\varphi(y_1 w_j) = \alpha_i$ for some i, j and we can obtain a $2n$-total-colouring of G as described in the earlier part of this proof. //

Lemma 6.19 Let $G = G(n_k, n_{k-1}, ..., n_1)$ be of order $2n + 1$. Suppose $n_k = n - 1$ and $n_{k-1} \geq 2$. Then G is Type 1.

Proof. We have $\bar{G} = S_{n_k} \cup ... \cup S_{n_1}$. Let the centre of $S_{n_k}, S_{n_{k-1}}, ..., S_{n_1}$ be w, $x_1, ..., x_{k-1}$ respectively.

Let z be an end-vertex of $S_{n_{k-1}}$. By Theorem 6.11, $G^* = G + x_1 z - wz$ has a $2n$-total-colouring φ. Let $m = 2(n+1-k)$. We shall modify φ to a $2n$-total-colouring of G using the same set $N = \{1, 2, ..., 2n\}$ of $2n$ colours.

By the proof of Theorem 6.11, we have

(i) the colours, each of which is used to colour exactly one vertex in $G - w$ excluding the end-vertices of $k - 1$ independent edges $x_1 y_1, ..., x_{k-1} y_{k-1}$ are $1, 2, ..., m$;

(ii) the colour used to colour the end-vertices of $x_i y_i$ is $m + i$, $i = 1, 2, ..., k - 1$;

(iii) the remaining $k - 1$ colours $m + k, ..., 2n$ were used to colour only edges of $G - w$, each such colour class forms a 1-factor of $G - w$;

(iv) the colours missing at the minor vertices of $G - w$ are the colours $1, 2, ..., m$, each colour is missing at only one vertex.

Let the end-vertices of S_{n_k} be $w_1, ..., w_{n-1}$ and let $w_n = z$. If $\varphi(x_1 w_i) = \beta \geq m + 1$ for some i, then we delete $x_1 w_i$ and put $\varphi(w w_i) = \beta$ to modify φ to a $2n$-total-colouring of G as described in the proof of Theorem 6.11. Hence we assume that $\beta_i = \varphi(x_1 w_i) \in M = \{1, 2, ..., m\}$ for all $i = 1, ..., n$.

Let $r = n_{k-1} - 1$ and let $\beta'_1, ..., \beta'_r$ be the colours missing at x_1. Then $\{\beta_1, ..., \beta_n\}$ and $\{\beta'_1, ..., \beta'_r\}$ are disjoint subsets of M. Hence there are $m - n - r = h$ other colours $\alpha_i \in M$, $i = 1, 2, ..., h$ used to colour some edges $x_1 u_i$, $i = 1, 2, ..., h$, where $\{u_1, ..., u_h\} \cap \{w_1, ..., w_n\} = \emptyset$. (Note that if $n + r > m$, then $\beta_i > m$ for some i, a contradiction to the fact that now $\beta_i \in M$ for all $i = 1, ..., n$.) We also know that there are $2n - 1 - m = t$ edges $x_1 z_j$, such that $\varphi(x_1 z_j) = \gamma_j > m$, $j = 1, ..., t$.

Let the set of colours missing at the vertices $z_1, ..., z_t$ be C. Since the missing colours at the minor vertices are in M and each colour in M is missing at only one minor vertex, C and $\{\beta'_1, ..., \beta'_r\}$ are disjoint subsets of M. If $C \cap \{\beta_1, ..., \beta_n\} \neq \emptyset$, say β_i is a colour missing at z_j, then we recolour the edge $x_1 z_j$ by β_i, delete the edge $x_1 w_i$, colour the edge $w w_i$ with β_i, colour the edge $w z_j$ with γ_j, and as before we can modify φ to a $2n$-total-colouring of G. Thus we can assume that $C \subseteq \{\alpha_1, ..., \alpha_h\}$ (if $t > h$, we obtain a contradiction here). By permutation of colours, we may further assume that the colours missing at $z_1, ..., z_t$ are $\alpha_1, ..., \alpha_{h_1}$.

Let the set of colours missing at $\{u_1, ..., u_{h_1}\}$ be C_1. If $C_1 \cap \{\beta_1, ..., \beta_n\} \neq \emptyset$, then we can apply the above "fan recolouring process" where the fan is pivoted at x_1 to modify φ to a $2n$-total-colouring of G. Hence we can assume that $C_1 \subseteq \{\alpha_1, ..., \alpha_h\}$.

Now C and C_1 are disjoint subsets of $\{\alpha_1, ..., \alpha_h\}$. By permutation of colours, we can assume that $C_1 = \{\alpha_{h_1+1}, \alpha_{h_1+2}, ..., \alpha_{h_2}\}$. We next consider the colours missing at the vertices $u_{h_1+1}, ..., u_{h_2}$ in a similar way. Since $\{\alpha_1, ..., \alpha_h\}$ is a finite set, eventually there is at least one colour β_i that will make the "fan recolouring process" work.   //

We shall make use of the following observation in the proof of Theorem 6.21.

Observation. In the proof of Lemma 6.19, the colours used to colour the vertices remain fixed while one of the colours from the set $\{m + k, ..., 2n\}$ is used to colour an edge wz.

Theorem 6.20 Let $G = G(n_k, n_{k-1}, ..., n_1)$ be of order $2n + 1$. Then G is Type 1 if one of the following holds:

(i) $n_k = n$ for $n - 1$;

(ii) $n_k = n_{k-1}$;

(iii) $n_k \leq n + 2 - k$.

Proof. (i) follows from Theorems 6.11, 6.18 and Lemma 6.19.

(ii) From $2(n_k + 1) + 2(k - 2) \leq (n_k + 1) + (n_{k-1} + 1) + ... + (n_1 + 1) = 2n + 1$, it follows that $2n_k \leq 2n - 2k + 3$. Hence $n_k \leq n - k + 1$. Now $e(\bar{G}) - \Delta(\bar{G}) = 2n + 1 - k - n_k \geq n$. Hence, by Theorem 6.14, G is Type 1.

(iii) Since $e(\bar{G}) - \Delta(\bar{G}) = 2n + 1 - k - n_k \geq n - 1$, by Theorem 6.14 and part (ii) above, G is Type 1.   //

From Theorem 6.20 it follows that $G = G(n_k, n_{k-1}, ..., n_1)$ is Type 1 for all $s \leq 1$ and $s \geq k - 2$, and consequently G is Type 1 for $k = 2, 3$. Also when $s = 2$ and $k = 4$, (iii) holds. Hence G is Type 1 when $k = 4$. Hence from now on we need only to consider the case that $k \geq 5$ and $k - 3 \geq s \geq 2$. We shall settle this remaining case by the following two theorems.

Theorem 6.21 Let $G = G(n_k, n_{k-1}, ..., n_1)$ be of order $2n + 1$. Suppose $n_k = n - s$, $k - 3 \geq s \geq 2$, and $2(k - 1) \leq n$. Then G is Type 1.

Proof. Let $w, x_1, ..., x_{k-1}$ be respectively the centres of the stars $S_{n_k}, S_{n_{k-1}}, ..., S_{n_1}$ in $\bar{G} = S_{n_k} \cup ... \cup S_{n_1}$.

We first note that $e(\overline{G-w}) - s = n - \alpha'(\overline{G-w}) = n - (k-1) \geq k-1$. Thus $\overline{G-w}$ has s edges $e_1 = x_1v_1,...,e_s = x^*v_s$ where x^* is the centre of some S_{n_j}, such that $G^* = G + e_1 + ... + e_s - wv_1 - ... - wv_s$ has the properties that $\alpha'(\overline{G-w}) = k-1$ and $e(\overline{G^* - w}) + \alpha'(\overline{G^* - w}) = n$ (in fact we choose $r_1 = n_{k-1} - 1$ edges $e_1,...,e_{r_1}$ from the end-vertices of $S_{n_{k-1}}$ etc). By Theorem 6.11, $G^* - w$ has a $2n$-total-colouring φ possessing the four properties given therein. By induction, we can assume that the edges $e_1 = x_1v_1,...,e_r = x'v_r$, $r \leq s-1$ (x' is the centre of some star S_{n_i}) had been put back into $\bar{G}^*$ and the edges $wv_1,...,wv_r$ had been coloured with r colours from the set $\{m+1,...,2n\}$, where $m = 2(n+1-k)$. We call this as a recolouring of $H = G^* - e_1 - ... - e_r + wv_1 + ... + wv_r$.

By induction we can assume that the set of colours involved in the recolouring of H is $S = \{m + j_1, ..., m + j_r\}$ and that each of the colours in $M \cup S$ is missing at exactly one minor vertex in $H - w$.

Let $\{\beta_1', ..., \beta_q'\}$ be the set of colours missing at x'' in the $2n$-total-colouring φ of H, where $e_{r+1} = x''v_{r+1}$ and x'' is the centre of some S_{n_j} in $\bar{G}$. Let $v_{r+1} = w_{n-s+1}, ..., v_s = w_{n-r}$ and let $\varphi(x''w_i) = \beta_i$, $i = 1, 2, ..., n-r$. If $\beta_i > m$ and $\beta_i \notin S$ for some $i = 1, 2, ..., n-r$, then we can delete the edge $x''w_i$, colour the edge ww_i by β_i to modify φ to a $2n$-total-colouring of $H - e_{r+1} + wv_{r+1}$. Hence we assume that $\{\beta_1, ..., \beta_{n-r}\} \subseteq M \cup S$.

Since $r \leq s-1 \leq k-4$, there are $p = 2n-1-m-r = 2n-1-2(n+1-k)-r = 2k-3-r \geq 2k-3-(k-4) = k+1$ edges $x''z_j$ such that $\varphi(x''z_j) = \gamma_j \notin M \cup S$, $j = 1, 2, ..., p$.

Let C be the set of colours missing at $z_1, z_2, ..., z_p$. Then $C \subseteq M$ and $C \cap \{\beta_1', ..., \beta_q'\} = \emptyset$. Since $d_{\overline{H}}(w) = n - r$ and the r edges $wv_1, ..., wv_r$ are edges in H, we know that there are at least

$$t = (n-r) - r = n - 2r \geq 2(k-1) - 2(k-4) = 6$$

colours in M which can be used to colour some edges e' incident with w in H such that $\varphi(e')$ was not the last colour used in a fan recolouring process where the fan is pivoted at some centre x of S_{n_i} ($xu_i = e_i$ for some $i \leq r$). Let this set of colours be C' and let $C_1 = \{\varphi(x''y) \in M; y \in N_H(x'')\}$. Since at most r colours from the set M had been used to colour some edges $wv_1, ..., wv_r$, and since $d_{\overline{H}}(x'') \leq q \leq n_k - 2$, we have

$$|C_1 \cap C'| \geq (m - q) - r \geq m - (n - s - 2) - r$$
$$\geq 2(n + 1 - k) - (n - s - 2) - (s - 1)$$
$$= n + 5 - 2k \geq 2(k - 1) + 5 - 2k = 3.$$

If $(C \cap C_1) \cap C' \neq \emptyset$, then using the "fan recolouring process" as described in the proof of Lemma 6.19, we can modify φ to a $2n$-total-colouring of $H + e_{r+1} - wv_{r+1}$. Hence we assume that $(C \cap C_1) \cap C' = \emptyset$. Let $C = \{\alpha_1, \alpha_2, ..., \alpha_c\}$, let V_C be the set of vertices v such that $\varphi(x''v) \in C$ and let $\{\delta_1, \delta_2, ...\}$ be the set of colours missing at V_C. Then $\{\delta_1, \delta_2, ..., \}$, C and $\{\beta'_1, ..., \beta'_q\}$ are disjoint subsets of M. If $(\{\delta_1, \delta_2, ...\} \cap C_1) \cap C' \neq \emptyset$, then applying the fan recolouring process where the fan is pivoted at x'', we can modify φ to a $2n$-total-colouring of $H + e_{r+1} - wv_{r+1}$. Otherwise continue considering the set of edges e' incident with x'' such that $\varphi(e') \in \{\delta_1, \delta_2, ...\}$ and applying the fan recolouring process and since M is a finite set, we will eventually obtain a colour in C' that can be used as a last colour to make the fan recolouring process work.   //

Applying Theorems 6.21 and 6.20 we can show that except for the two graphs $G(4, 1, 1, 1, 1)$ and $G(5, 2, 1, 1, 1)$, any graph $G = G(n_5, n_4, ..., n_1)$ of order $2n + 1$ is Type 1 (see Exercise 6(9)). We now show that $G = G(4, 1, 1, 1, 1)$ is Type 1. The technique for proving that $G(4, 1, 1, 1, 1)$ is Type 1 will be applied to settle the remaining case that $n_k = n - s$, $k - 3 \geq s \geq 2$ and $n < 2(k - 1)$.

Example 6.7 $G = G(4, 1, 1, 1, 1)$ is Type 1.

We refer to the graph $\bar{G}$ given in Fig. 6.6. Let $G^* = G + x_1 y_1 - wx_1 - wy_1$.

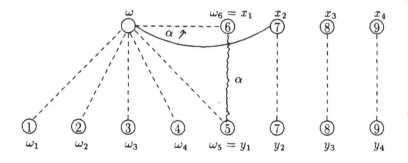

Figure 6.6

Then $e(\overline{G^* - w}) + \alpha'(\overline{G^* - w}) = 3 + 3 = 6$. Hence, by Theorem 6.11, G^* has a 12-total-colouring φ in which its vertices receive the colours as shown in Fig. 6.6. Refering to $G^* = G(6,1,1,1)$ we have $k = 4$, and $m = 2(n + 1 - k) = 6$.

Case (i) Suppose $\varphi(w_i w_j) = \alpha \geq 10 > 9 = m + k - 1$ for some i, j, say $\varphi(w_5 w_6) = \alpha$. Then by deleting $w_5 w_6$ from G^*, setting $\varphi(w x_1) = \alpha$, recolouring x_2 by colour 5 (assuming that colour 5 is missing at x_2), recolouring y_1 by colour α, and setting $\varphi(w y_1) = 5$ and $\varphi(w x_2) = 7$, we obtain a 12-total-colouring of G.

Case (ii) Suppose $\varphi(w_i w_j) < 10$ for any $i \neq j$. Since $K_6 = G^*[w_1, w_2, ..., w_6]$ cannot be edge-coloured with only three colours 7, 8 and 9, we have $\varphi(w_i w_j) \leq 6$ for some $i \neq j$. Let $\varphi(w_5 w_6) = \beta(\leq 4)$. Let $\{z_1, ..., z_6\} = \{x_2, x_3, x_4, y_2, y_3, y_4\}$. Let P be the $(\beta, 10)$-path (the edges of this path are alternately coloured with colours β and 10) containing $x_1 y_1 = w_6 w_5$. (If it is a cycle C, then by interchanging the colours β and 10 in C, it reduces to Case 1.) One terminus of P must be a vertex in $\{z_1, ..., z_6\}$ say z_2, where colour β is missing. The other terminus of P is a vertex w_i such that $\varphi(w_i) = \beta$.

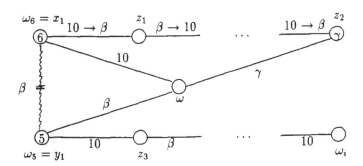

(β is absent at z_2; γ is replaced by 10 at z_2)

Figure 6.7

Suppose the situation is given as in Figure 6.7. (By symmetry, this is a general situation.) By deleting the edge $x_1 y_1$ from G^*, interchanging the colours β and 10 in the subpath $x_1 z_1 ... z_2$, (see Fig. 6.7), setting $\varphi(z_2) = 10$, $\varphi(w x_1) = 10$, $\varphi(w y_1) = \beta$, and $\varphi(wz_2) = \gamma$ where γ is the colour originally used to colour vertex z_2, we can modify φ to a 12-total-colouring of G. //

The above method can also be used to show that $G(5,2,1,1,1)$ is Type 1.

Remarks. We observe that in both Case (i) and Case (ii) in Example 6.7, only one of the colours in $\{10,11,12\}$ is used to replace a colour in $\{1,2,..,9\}$ in the vertex-colouring. We shall make use of this fact in the proof of Theorem 6.22.

Theorem 6.22 Let $G = G(n_k, n_{k-1}, ..., n_1)$ be of order $2n + 1$. Suppose $n_k = n - s$, $k - 3 \geq s \geq 2$ and $n < 2(k-1)$. Then G is Type 1.

Proof. We first prove that $\overline{G}$ contains at least one S_1. Suppose otherwise. Then

$$(n_k + 1) + 3(k - 1) \leq 2n + 1$$

implies that $(n-s+1)+3(k-1) \leq 2n+1$. Hence $3(k-1) \leq n+s < 2(k-1)+(k-3) = 3k - 5$, which is false.

Now let the edge of S_{n_1} in $\bar{G}$ be $a_1 b_1$ and let $G_1 = G + a_1 b_1 - wa_1 - wb_1$, where w is the centre of S_{n_k} in $\bar{G}$. Then

$$G_1 = G(n_k + 2, n_{k-1}, ..., n_2) = G(n'_{k-1}, n'_{k-2}, ..., n'_1).$$

Let $k' = k - 1$ and $s' = s - 2$. We observe that $k' - 3 \geq s' \geq 0$. We consider two cases separately.

Case 1. $n \geq 2(k' - 1)$.

In this case we let φ be a $2n$-total-colouring of G_1 constructed by the proof of Theorem 6.21. By permutation of colours we can assume that the colours used to colour the s' edges $wu_1, ..., wu_{s'}$ are $m + j_1, ..., m + j_{s'}$, where $m = 2(n + 1 - k') = 2(n + 2 - k)$. Let $S' = \{m + j_1, ..., m + j_{s'}\}$.

Subcase 1(i). Suppose $\varphi(w_i w_j) = \alpha \notin M \cup S'$ for some $i, j \leq n+2-s'$ (here we use the notation of the proof of Theorem 6.21). Then we can apply the proof technique used in Case (i) of Example 6.7 to show that φ can be modified to a $2n$-total-colouring of G.

Subcase 1(ii). Suppose $\varphi(w_i w_j) \in M \cup S'$ for all i, $j \leq n + 2 - s'$. Let $\varphi(w_{n+1-s'} w_{n+2-s'}) = \beta$ and let $\gamma \in \{m + 1, ..., 2n\} \setminus S'$. Let P be the (β, γ)-path containing the edge $w_{n+1-s'} w_{n+2-s'}$. Then applying the proof technique used in Case (ii) of Example 6.7 we can also show that φ can be modified to a $2n$-total-colouring of G.

Case 2. $n < 2(k' - 1)$.

We shall settle this case by induction on k. Let q be the number of copies of S_1 in $\bar{G}$ such that $n_k + 2q \leq n$, but $n \geq 2(k - q - 1)$ (the existence of q follows by induction on k and $n < 2(k - (q - 1) - 1) = 2(k - q)$) or $n - 1 \leq n_k + 2q \leq n$ and $n < 2(k - q - 1)$. Let $a_1 b_1, ..., a_q b_q$ be the q copies of S_1 in $\bar{G}$, and let $G^* = G + (a_1 b_1 + ... + a_{q-1} b_{q-1}) - w a_1 - w b_1 - ... - w a_q - w b_q$. By induction, G^* has a $2n$-total-colouring φ which, by the proof technique of Example 6.7, can be modified to a $2n$-total-colouring of $G' = G + a_q b_q - w a_q - w b_q$ for which φ has the property that each of the $(k - 1) - q = k - 1 - q$ colours $m' + 1, ..., m' + k - 1 - q$ $(m' = 2n - 2(k - 1 - q) = 2(n + 1 - k + q))$ was used to colour exactly two end-vertices of $k - 1 - q$ independent edges in $\overline{G^*}$ and (by permutation of colours if necessary) that each of the $q - 1$ colours $m' + k - q, ..., m' + k - 2$ was used exactly once in recolouring one of the end-vertices of $a_1 b_1, ..., a_{q-1} b_{q-1}$ in G^*. Now $2n - (m' + k - 2) = k - 2q$ and $k - 2 \geq n - k + 3 \geq 3$ in the later case that $n - 1 \leq n_k + 2q \leq n$ and $n < 2(k - q - 1)$. Also in the former case that $n_k + 2p \leq n$ but $n \geq 2(k - q - 1)$, we have $k - 2q \geq n + 1 - k \geq 2$ (otherwise $n + 1 - k \leq 1$ implies that $n \leq k$ and so $n = k$ and $G = G(2, 1, 1, ..., 1)$ which in turn implies that $n = n_k + s = 2 + s \leq 2 + (k - 3) = k - 1$, which is false). Thus there are always at least one colour $\gamma \geq m' + k - q$ which can be used to recolour a vertex w_i in the recolouring process. Hence we can apply the proof technique of Example 6.7 to modify φ to a $2n$-total-colouring of G. //

Combining all the results of Theorem 6.10 to Theorem 6.23, we have

Theorem 6.24 (Yap, Chen and Fu [-a]) Suppose G is a graph of odd order $2n + 1$ having $\Delta(G) = 2n - 1$. Let w be a vertex of minimum degree in G. Then G is Type 1 if and only if

$$e(\overline{G - w}) + \alpha'(\overline{G - w}) \geq n.$$

§4. Further results

In this section we shall determine $\chi_T(G)$ for the following classes of graphs:

(i) Nearly complete bipartite graphs G;
(ii) Graphs G such that $\bar{G}$ is bipartite;
(iii) Some graphs G having $\Delta(G) = |G| - 3$.

Hilton [91] gives a classification of nearly complete bipartite graphs according to their total chromatic numbers. As this chapter is getting too long, we shall only state it here.

Theorem 6.25 (Hilton [91]). Let $J \subseteq K_{n,n}$. Then

$$\chi_T(K_{n,n} - E(J)) = n + 2$$

if and only if $e(J) + \alpha'(J) \leq n - 1$.

From Theorem 6.25, we have (see Exercise 6(6)),

Corollary 6.26 Let $J \subseteq K_{n,n}$ and let $G = K_{n,n} - E(J)$. Then

$$\chi_T(G) = \begin{cases} \Delta(G) + 2 & \text{if } e(J) + \alpha'(J) \leq n - 1, \\ \Delta(G) + 1 & \text{if } 2n - 1 \geq e(J) + \alpha'(J) \geq n. \end{cases}$$

The necessity part of Theorem 6.25 has been generalized by Yap [95]. The proof of this generalized result is almost identical to that of Hilton's original proof.

Theorem 6.27 (Hilton [91]; Yap [95]) Let $J \subseteq K_{n,n}$. If

$$\chi_T(K_{n,n} - E(J)) = t,$$

then $e(J) + \alpha'(J) \geq n(n + 2 - t)$.

Proof. Let $G = K_{n,n} - E(J)$. Let φ be a total-colouring of G using t colours $c_1, c_2, ..., c_t$. Let (X, Y) be the bipartition of $K_{n,n}$. Let $X_i \cup Y_i$, $i = 1, ..., t$ ($X_i \subseteq X$, $Y_i \subseteq Y$) be the vertex-colour classes of φ, and let

$$m_i = \min\{|X_i|, |Y_i|\}, \quad f_i = \max\{|X_i|, |Y_i|\} - m_i.$$

Then

$$|X_i| + |Y_i| = f_i + 2m_i.$$

Hence

$$(f_1 + f_2 + ... + f_t) + 2(m_1 + m_2 + ... + m_t) = 2n.$$

Since $X_i \cup Y_i$ is an independent set of vertices in G, there are m_i independent edges in $J[X_i \cup Y_i]$. Thus

$$m \geq m_1 + m_2 + ... + m_t,$$

and so

(7) $$f_1 + f_2 + ... + f_t \geq 2n - 2m.$$

Let $C = \{c_1, c_2, ..., c_t\}$. For each $c \in C$ and each $v \in V(G)$, we call (c, v) a colour-vertex pair if c is present at v, i.e., either $\varphi(v) = c$ or $\varphi(vu) = c$ for some edge vu.

We note that if $|X_i| \geq |Y_i|$, say, then $f_i = |X_i| - |Y_i|$ and thus there are at least f_i vertices in Y in which colour c_i is absent. Hence there are at most $2n - f_i$ colour-vertex pairs (c_i, v), $v \in V(G)$, and so there are at most $2nt - (f_1 + f_2 + ... + f_t)$ colour-vertex pairs altogether. But the number of colour-vertex pairs equals the number of vertices plus twice the number of edges, and so is $2n + 2(n^2 - e(J))$. Thus, by (7), we have

$$2n + 2(n^2 - e(J)) \leq 2nt - (f_1 + f_2 + ... + f_t) \leq 2nt - (2n - 2\alpha'(J)).$$

Consequently, $e(J) + \alpha'(J) \geq n(n + 2 - t)$.   //

Corollary 6.28 Let $J \subseteq K_{n,n}$. If $G = K_{n,n} - E(J)$ is Type 1, then $e(J) + \alpha'(J) \geq n(n + 1 - \Delta(G))$.

We note that the converse of Corollary 6.28 is in general not true. For example, we take $J = C_8 \subseteq K_{4,4}$. Then $G = K_{4,4} - E(J) \simeq C_8$. But $\chi_T(C_8) = 4 \neq \Delta(C_8) + 1$.

Dugdale and Hilton [90] determine $\chi_T(G)$ for the class of d-regular graphs G such that $\bar{G}$ is bipartite (Theorem 6.34). They apply the following lemma to prove

Lemma 6.30 which together with Lemma 6.31 (König's theorem; see Yap [86;p.11]), Lemma 6.32 and Lemma 6.33 are used to prove Theorem 6.34. Note that Lemma 6.32 is a special case of a result due to Ghouila-Houri (1960), for a proof, see J. A. Bondy and U. S. R. Murty : Graph Theory with Applications, p.178.

We need the following definitions in Lemma 6.29. We call a latin square of side n on the symbols $\sigma_1, \sigma_2, ..., \sigma_n$ symmetric if the two cells (i, j) and (j, i) contain the same symbol for any $1 \leq i \leq j \leq n$. If the symbol in cell (i, i) is σ_i for all $i = 1, 2, ..., n$, then we say that the latin square is idempotent.

Lemma 6.29 Let $n \geq 3$ be an odd integer, $n = 2m+1$, and let $K_{n,n}$ have bipartition (X, Y), where $X = \{x_1, ..., x_n\}$ and $Y = \{y_1, ..., y_n\}$. Then there exists an edge-colouring ϕ of $K_{n,n}$ using colours $1, ..., n$, such that

(i) $\phi(x_i y_i) = i$ for $1 \leq i \leq n$;
(ii) $\phi(x_{2k} y_{2k+1}) = \phi(x_{2k+1} y_{2k}) = 1$ for $1 \leq k \leq m$.

Proof. It is well-known that there exists a symmetric idempotent latin square $L = (l_{ij})$ of side n with symbols $1, 2, \cdots, n$ (using the natural colouring of K_n we can obtain this result) such that $l_{ii} = i$ for $i = 1, 2, ..., n$. If $l_{23} = l_{32} = 1$, we let the second and third row (column) of L remain as they are. If $l_{2j} = l_{j2} = 1$, $j \geq 4$, then we permute the 3rd column with the jth column and simultaneously we permute the 3rd row with the jth row. We next interchange the symbol 3 with the symbol l_{jj} throughout the whole latin square. We now have a new symmetric idempotent latin square $L^{(1)} = (l_{ij}^{(1)})$ with symbols $1, 2, \cdots, n$ such that $l_{ii}^{(1)} = i$, $i = 1, ..., n$ and $l_{23}^{(1)} = l_{32}^{(1)} = 1$. We perform the same process on the fourth row (column) and fifth row (column) and so on. Eventually we will obtain a symmetric idempotent latin square $L' = (l_{ij}')$ such that $l_{ii}' = i$, $i = 1, 2, \cdots, n$ and $l_{2k,2k+1}' = l_{2k+1,2k}' = 1$ for all $k = 1, 2, \cdots, m$. This situation is depicted in Fig. 6.8.

Finally, we put $\phi(x_i y_j) = l_{ij}'$ to obtain a required n-edge-colouring of $K_{n,n}$.   //

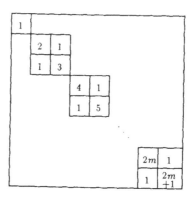

Figure 6.8

Lemma 6.30 For any integer $n \geq 4$, there exists an $(2n-1)$-edge-colouring of K_{2n} with a Hamilton path receiving distinct colours.

Proof. We split the vertex set of $G = K_{2n}$ into disjoint union of $X = \{x_1, x_2, \ldots, x_n\}$ and $Y = \{y_1, y_2, \ldots, y_n\}$. For convenience we again call $G[X]$ and $G[Y]$ the upper K_n and the lower K_n, which are denoted by G_1 and G_2 respectively. We let H be the complete bipartite graph $G(X, Y)$. We will prove this lemma by induction on n.

Case 1. $n \geq 4$ is even.

For $n = 4$, in the natural edge-colouring of K_8, the edges of the Hamilton path $v_1 v_2 v_8 v_4 v_3 v_6 v_7 v_5$ receive distinct colours $5, 2, 4, 7, 1, 3$ and 6. For $n = 6$, in the natural edge-colouring of K_{12}, the edges of the Hamilton path $v_9 v_{11} v_8 v_5 v_6 v_{12} v_2 v_1 v_4 v_3 v_7 v_{10}$ receive distinct colours $10, 4, 1, 11, 6, 2, 7, 8, 9, 5$ and 3.

For $n \geq 8$, by the induction hypothesis, G_1 (resp. G_2) has an edge-colouring using colours $n+1, n+2, \ldots, 2n-1$ such that the Hamilton path $x_1 x_2 \ldots x_n$ (resp. $y_1 y_2 \ldots y_n$) receive distinct colours $n+1, n+2, \ldots, 2n-1$.

By Lemma 3.3 we know that H has an edge-colouring using colours $1, 2, \ldots, n$ such that the edges $x_1 y_1, x_2 y_2, \ldots, x_n y_n$ receive distinct colours $1, 2, \ldots, n$. We can now combine these edge-colourings to obtain a required edge-colouring of K_{2n} such that the Hamilton path $x_1 y_1 y_2 x_2 x_3 y_3 \ldots x_{n-1} y_{n-1} y_n x_n$ receive distinct colours

$1, n+1, 2, n+2, 3, ..., n-1, 2n-1, n$ as depicted in Fig. 6.9.

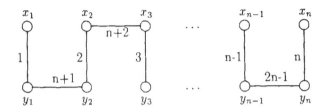

Figure 6.9

Case 2. $n \geq 5$ is odd.

For $n = 5$, we exhibit in Fig.6.10 a latin square corresponding to an edge-colouring of K_{10}: Cell (i, j) is filled with integer k if and only if the edge $v_i v_j$ is coloured with colour k. The shaded cells corresponding to the edges of a Hamilton path receive distinct colours.

	v_1	v_2	v_3	v_4	v_5	v_6	v_7	v_8	v_9	v_{10}
v_1		4	5	2	3	6	8	1	7	9
v_2	4		3	5	1	2	7	9	6	8
v_3	5	3		1	2	9	6	8	4	7
v_4	2	5	1		4	8	3	7	9	6
v_5	3	1	2	4		7	9	6	8	5
v_6	6	2	9	8	7		1	5	3	4
v_7	8	7	6	3	9	1		4	5	2
v_8	1	9	8	7	6	5	4		2	3
v_9	7	6	4	9	8	3	5	2		1
v_{10}	9	8	7	6	5	4	2	3	1	

Figure 6.10

For $n = 2m + 1 \geq 7$, we add a vertex y_0 to the vertex set of G_2. Let G_2' be the complete graph K_{n+1} with vertex set $\{y_0, y_1, \ldots, y_n\}$. By induction, we can edge-colour G_2' using n colours $1, n+1, \ldots, 2n-1$ such that the edges of the Hamilton path $y_0 y_1 y_2 \ldots y_n$ receive distinct colours $1, n+1, n+2, \ldots, 2n-1$. In G_1, denote $z_1 = x_1$, and for $1 \le k \le m$ denote $z_{2k} = x_{2k+1}$ and $z_{2k+1} = x_{2k}$. Then add a vertex z_0 to vertex set $\{z_1, \ldots, z_n\}$. Colour the edge $z_i z_j$ of the complete graph having vertex set $\{z_0, z_1, \ldots, z_n\}$ with the same colour as that of the edge $y_i y_j$ in G_2' for all $0 \le i < j \le n$.

Now delete the vertices y_0 and z_0. Then the vertex y_i has the same missing colour as z_i ($2 \le i \le n$). Recolour the edge $y_i z_i$ with this missing colour. By Lemma 6.29, the edges $x_{2k} y_{2k+1}$ and $x_{2k+1} y_{2k}$, $1 \le k \le m$, which were originally coloured with colour 1, have now been recoloured with the respective missing colours at y_i and z_i. Also, the edge $x_i y_i$ is coloured with colour i, $i = 2, \ldots, n$. We obtain an edge-colouring of K_{2n} such that the edges of the Hamilton path $x_1 y_1 y_2 x_2 x_3 y_3 \ldots y_{n-1} x_{n-1} x_n y_n$ receive distinct colours $1, n+1, 2, n+2, 3, n+3, \ldots, n-1, 2n-1$ and n as depicted in Fig. 6.11. //

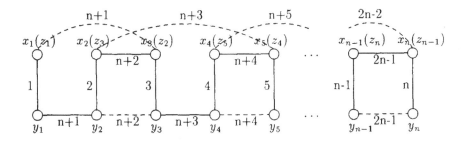

Figure 6.11

Lemma 6.31 (König's theorem) A regular bipartite graph of degree d is the union of d edge-disjoint 1-factors.

Lemma 6.32 (Ghouila-Houri) Let D be a directed graph of order $v \ge 3$. Suppose the in-degree $d^-(x)$ and out-degree $d^+(x)$ of each vertex x of D satisfy

$$d^-(x) \ge \frac{1}{2}v, \quad d^+(x) \ge \frac{1}{2}v.$$

Then D contains a directed Hamilton cycle.

Lemma 6.33 Let B be an m-regular bipartite graph of order $4m$. Then either B contains a Hamilton cycle, or B consists of two disjoint copies of $K_{m,m}$.

(This lemma is due to A. G. Chetwynd and R. Häggkvist, as well as P. Ash. For details of source of references, refer to Dugdale and Hilton [94].)

Theorem 6.34 (Dugdale and Hilton [94]) A d-regular graph G of even order $2n \geq 6$ whose complement $\bar{G}$ is bipartite has total chromatic number $d + 1$ if and only if G is not a complete graph, and $\bar{G} \neq K_{n,n}$ when n is even.

Proof. Let G be a d-regular graph of even order $2n \geq 6$ such that $\bar{G}$ is bipartite. If $G = K_{2n}$ or $K_{n,n}$ where n is even, then by Theorem 3.1, G is Type 2. Hence, from now on we assume that $G \neq K_{2n}$ and if n is even, then $\bar{G} \neq K_{n,n}$. We shall then show that $\chi_T(G) = d + 1$.

Since G is regular and $\bar{G}$ is bipartite, G is the union of two vertex-disjoint K_n's which we denote by G_1 and G_2. Let $V(G_1) = X = \{x_1, ..., x_n\}$ and $V(G_2) = Y = \{y_1, ..., y_n\}$. Then G contains a regular bipartite subgraph graph $B \subseteq G(X, Y)$ of degree $d - n + 1$. Note that $\bar{G}$ is regular of degree $2n - 1 - d \geq 1$.

We consider two cases.

Case 1. n is odd.

By König's theorem, $\bar{G}$ contains a 1-factor, say $\{x_1 y_1, ..., x_n y_n\}$. Using the natural edge-colouring, G_1 and G_2 can be edge coloured with n colours $1, 2, ..., n$ such that colour i is missing at vertices x_i and y_i. We use colour i to colour the vertices x_i and y_i, $i = 1, ..., n$. We then use $d - n + 1$ colours $n + 1, ..., d + 1$ to colour the edges of B. Thus $\chi_T(G) \leq d + 1$ and consequently $\chi_T(G) = d + 1$.

Case 2. n is even.

In this case $n \geq 4$ and since $G \neq K_{n,n}$, the degree of B is nonzero. We shall suppose first that $n \geq 8$. We consider three subcases.

Case 2a. $d \leq \frac{1}{2}(3n - 4)$.

By König's theorem, B contains a 1-factor $F_1 = \{x_1 y_1, ..., x_n y_n\}$ say. We construct a directed graph D on the vertices $y_1, ..., y_n$ as follows. For $i \leq j \leq n$, we put

$$\overrightarrow{y_i y_j} \in A(D) \quad \text{if} \quad x_{i+1} y_j \notin E(G).$$

($A(D)$ is the arc set of D and the subscripts are to be read mod n.) Let $d^-(D)$ and $d^+(D)$ denote, respectively, the minimum in and out degrees of the vertices of D. Since $\bar{G}$ is regular of degree $2n - 1 - d$ and since an edge $x_{i+1}y_i$ of $\bar{G}$ does not give rise to an arc in D, it follows that

$$\min\{d^-(D), d^+(D)\} \geq 2n - 1 - d - 1 = 2n - d - 2.$$

But $d \leq (3n - 4)/2$ and so

$$\min\{d^-(D), d^+(D)\} \geq 2n - (3n - 4)/2 - 2 = \frac{n}{2}.$$

Therefore, by Lemma 6.32, D contains a directed Hamilton cycle C. For $1 \leq i \leq n$ let $y_i^* \in \{y_1, y_2, ..., y_n\}$ be such that $\overrightarrow{y_i y_i^*}$ is an arc in C. Thus C consists of the arcs $\overrightarrow{y_1 y_1^*}, \overrightarrow{y_2 y_2^*}, ..., \overrightarrow{y_n y_n^*}$ in some order. Moreover, $y_1^*, y_2^*, ..., y_n^*$ are such that, for $1 \leq i \leq n$, $x_{i+1}y_i^*$ is an edge in $\bar{G}$, $y_i \neq y_i^*$ and $\{y_i y_i^* | i \leq i \leq n\}$ forms a Hamilton cycle in G_2. Let F_2 be the 1-factor of $\bar{G}$ consisting of $\{x_{i+1}y_i^* | 1 \leq i \leq n\}$.

For $1 \leq i \leq n - 1$, colour $x_i x_{i+1}$ with colour c_i. Then by Lemma 6.30, since $n \geq 8$, this partial edge-colouring can be extended to a complete edge-colouring of G_1 using only the colours $c_1, c_2, ..., c_{n-1}$. Now for $1 \leq i \leq n - 1$ remove the colour c_i from $x_i x_{i+1}$ and colour the vertex x_{i+1} with c_i.

For $1 \leq i \leq n - 1$, colour $y_i y_i^*$ with colour c_i. Then by Lemma 6.30 this partial colouring can be extended to an edge colouring of G_2 using only the colours $c_1, c_2, ..., c_{n-1}$. Remove the colour c_i from $y_i y_i^*$ and colour the vertex y_i^* with colour c_i.

For $1 \leq i \leq n - 1$, colour the edge $x_i y_i$ with colour c_i.

Next consider the path $P : x_1 x_2 ... x_n y_n ... y_n^*$ where $y_n ... y_n^*$ is the reverse of a directed Hamilton path in G_2. Note that all the vertices of P except x_1 and y_n^* have been coloured but none of the edges have been coloured; also $x_1 y_n^* = x_{n+1} y_n^*$ is in $\bar{G}$. Now colour the vertices x_1 and y_n^* with colour c_n and the edges of P with colours c_{n+1} and c_n alternately.

The edges of G which are not yet coloured all lie in B and by König's theorem can be coloured with $d - n$ colours $c_{n+2}, c_{n+3}, ..., c_{d+1}$.

Thus G has a total-colouring using $d + 1$ colours.

Case 2b. $d \geq \frac{3}{2}n$.

By König's theorem, $\bar{G}$ contains a 1-factor F_2, say.

We may suppose this time that $V(G_1) = X$ and $V(G_2) = \{y_1^*, ..., y_n^*\}$, where, for $1 \leq i \leq n$, the edge $x_{i+1}y_i^* \in F_2$.

Now construct a directed graph D' on the vertices $y_1^*, y_2^*, ..., y_n^*$ as follows. For $1 \leq i \leq n$, put $\overrightarrow{y_j^* y_i^*} \in A(D')$ if $j \neq i$ and $x_i y_j^* \in E(G)$. Since B is regular of degree $d - (n - 1)$ and since $x_i y_i^* \in E(G)$ does not give rise to an arc in D', it follows that

$$\min\{d^-(D'), d^+(D')\} \geq d - (n - 1) - 1 = d - n.$$

Since $d \geq \frac{3}{2}n$, it follows that

$$\min\{d^-(D'), d^+(D')\} \geq \frac{3}{2}n - n = \frac{n}{2}.$$

Hence by Lemma 6.32, D' contains a directed Hamilton cycle C. For $1 \leq i \leq n$, let $y_i \in \{y_1^*, y_2^*, ..., y_n^*\}$ be such that $\overrightarrow{y_i y_i^*}$ is an arc of C. Thus C consists of the arcs $\overrightarrow{y_1 y_1^*}, \overrightarrow{y_2 y_2^*}, ..., \overrightarrow{y_n y_n^*}$ in some order. Moreover, $y_1, y_2, ..., y_n$ are such that $x_i y_i \in E(G)$, for all $i = 1, 2, ..., n$. Let $F_1 = \{x_1 y_1, x_2 y_2, ..., x_n y_n\}$. The argument now proceeds as in Case 2a.

Case 2c. $d = \frac{1}{2}(3n - 2)$.

In this case B is regular of degree $d - (n - 1) = \frac{1}{2}(3n - 2) - (n - 1) = \frac{n}{2}$. By Lemma 3.33, either B has a Hamilton cycle C or B consists of two copies of $K_{\frac{n}{2},\frac{n}{2}}$.

Suppose first that B has a Hamilton cycle $C : x_1 y_1 x_2 y_2 x_3 ... x_n y_n x_1$. The proof now proceeds as in Case 2a except that, for $1 \leq i \leq n$, $x_{i+1} y_i \in E(G)$ and so

$$\min\{d^-(D), d^+(D)\} \geq 2n - 1 - d = \frac{n}{2}.$$

Thus we can apply Lemma 6.32 to show that G is Type 1 as in Case 2a.

Next suppose that B consists of two copies of $K_{\frac{n}{2},\frac{n}{2}}$. Assume that one copy of $K_{\frac{n}{2},\frac{n}{2}}$ has bipartition $(\{x_1, ..., x_{\frac{n}{2}}\}, \{y_1, ..., y_{\frac{n}{2}}\})$. Clearly B contains a 1-factor $F_1 = \{x_1 y_1, ..., x_n y_n\}$. We construct the directed graph D as in Case 2a. It contains the directed Hamilton cycle $y_1 y_{\frac{n}{2}+1} y_2 y_{\frac{n}{2}+2} \cdots y_{\frac{n}{2}-3} y_{n-3} y_{\frac{n}{2}-2} y_n y_{n-1} y_{\frac{n}{2}} y_{\frac{n}{2}-1} y_{n-2} y_1$. The proof now proceeds as in Case 2a.

Next we settle the case $n = 4$. Clearly in this case $1 \leq d(B) \leq 3$ and we can label the vertices $X = \{x_1, x_2, x_3, x_4\}$ and $Y = \{y_1, y_2, y_3, y_4\}$ in such a way that $x_i y_i \in E(B)$ and $x_i y_{i+2} \notin E(B)$. For $1 \leq i \leq 4$, colour vertices x_i, y_{i+2} and edges $x_{i+1} x_{i+2}, y_i y_{i+1}, x_{i+3} y_{i+3}$ with colour c_i (the subscript are to be read mod 4), and colour the edges $x_1 x_3, x_2 x_4, y_1 y_3, y_2 y_4$ with c_5. The remaining edges form a regular bipartite graph of degree $d - 4$ and can be coloured with colours $c_6, ..., c_{d+1}$. This gives a total-colouring of G with $d + 1$ colours.

Finally, we settle the case $n = 6$. In this case $1 \leq d(B) \leq 5$ and it is not difficult to see that we can label the vertices $X \cup Y$ such that $x_i y_i \in E(B)$ and $x_i y_{i+1} \notin E(B)$, $1 \leq i \leq 6$. For $i = 1, 3, 5$, colour the following vertices and edges with c_i : $x_{i+4}, y_{i+5}, x_i y_i, x_{i+1} x_{i+2}, x_{i+3} x_{i+5}, y_{i+1} y_{i+3}, y_{i+2} y_{i+4}$; for $i = 2, 4, 6$ colour the following vertices and edges with c_i : $x_{i+4}, y_{i+5}, x_i y_i, x_{i+1} x_{i+2}, x_{i+3} x_{i+5}, y_{i+2} y_{i+3}, y_{i+1} y_{i+4}$; and colour the following edges with c_7 : $x_1 x_4, x_2 x_5, x_3 x_6, y_1 y_2, y_3 y_4, y_5 y_6$ (the subscripts are to be read mod 6). The remaining edges of G form a regular bipartite graph of degree $d - 6$, and so can be coloured with $c_8, ..., c_{d+1}$. This gives a total-colouring of G with $d + 1$ colours. //

(Remarks. Theorem 6.34 has been generalized by Hilton and Zhao [-a].)

To end this chapter, we mention some other results on the classification of graphs according to their total chromatic number. Becuase of the space constraint of this chapter, we shall omit their proofs.

The following are some partial results on the classification of graphs G having $\Delta(G) = |G| - 3$ according to their total chromatic numbers.

Theorem 6.35 (Dugdale and Hilton [90]) Let G be a graph of even order $2n$. Suppose G is $(2n - 3)$-regular. Then G is Type 1 if and only if $\bar{G}$ is the union of even cycles.

Theorem 6.36 (Yap [93]) Let G be a graph of even order having $\Delta(G) = |G| - 3 \geq 3$. Suppose $\bar{G}$ is the union of two odd cycles plus an edge. Then G is Type 1.

Remarks. (i) From Theorem 2.7 and Theorem 6.36, it follows that if G is of even order and $\bar{G}$ is the union of two odd cycles, then G is total-chromatic critical.

(ii) Chetwynd and Hilton had proved that if G is a graph of odd order $2n+1$ and G is $(2n-2)$-regular, then G is Type 1 if and only if $\bar{G} \geq K_3$ (see also Yap [93]).

Theorem 6.37 (Yap [93]) Let G be a graph of odd order $|G| \geq 5$ having $\Delta(G) = |G| - 3$. Suppose the deficiency of G is 2. Then G is Type 1.

Theorem 6.38 (Chetwynd, Hilton and Zhao [91]) Suppose G is a k-regular graph of odd order. If $k \geq \frac{\sqrt{7}}{3}|G|$, then G is Type 1 if and only if every subgraph H of G with $\Delta(H) = \Delta(G)$ is conformable.

For the case that $|G| = 2n+1$ and $k = 2n-4$, Theorem 6.38 says that G is Type 1 if $|G| \geq 43$. The following result reduces the lower bound of $|G|$ to 21.

Theorem 3.39 (Yap and Chu [-a]) Let G be a connected graph of odd order $2n+1 \geq 21$. Suppose G is $(2n-4)$-regular. Then G is Type 1 if and only if G contains O_5 or G contains $O_3 \cup O_3$.

Remarks. From the result of Chetwynd-Hilton mentioned in Remark (ii) above and Theorem 6.37 it follows that if G is of odd order and $\bar{G}$ is the union of two cycles, and $\bar{G} \not\geq K_3$ then G is total-chromatic critical.

Exercise 6

1. Suppose G is a graph of order $2n$. Prove that if $2n-1 \geq e(\bar{G}) + \alpha'(\bar{G}) \geq n$, then $\Delta(G) = 2n-1$ (and thus by Theorem 6.1, G is Type 1). (Hilton [89/90])

2. Let G be a graph of order 8 and let $\bar{G} = C_3 \cup O_5$. Following the algorithm given in the proof of Theorem 6.1, find a total-colouring of G using 8 colours.

3. Let the bipartition of $K_{4,4}$ be (X,Y) and $S = S_2 \cup 2S_1 \subseteq K_{4,4}$. Find a 4-edge-colouring of $K_{4,4}$ such that the four edges of S receive distinct colours.

4. For each of the following spanning forest S of K_{10}, find a 9-edge-colouring φ of K_{10} such that all the edges of S receive distinct colours.

   (i) $S = S_5 \cup 2S_1$ (ii) $S = S_4 \cup S_2 \cup S_1$ (iii) $S = 2S_3 \cup S_1$
   (iv) $S = S_3 \cup 2S_2$ (v) $S = S_3 \cup 3S_1$ (vi) $S = 2S_2 \cup 2S_1$.

5. For each of the following spanning forest S of K_{12}, find an 11-edge-colouring of K_{12} such that all the edges of S receive distinct colours.

   (i) $S = S_7 \cup 2S_1$ (ii) $S = S_5 \cup 2S_2$ (iii) $S = S_4 \cup S_3 \cup S_2$ (iv) $S = 3S_3$.

6. Deduce Corollary 6.26 from Theorem 6.25.

7. Suppose $J \subseteq K_{n,n}$, $G = K_{n,n} - E(J)$ and $\Delta(G) = n - 1$. If J contains a 1-factor of $K_{n,n}$, prove that G is Type 1. (Hint: Apply Lemma 3.3 and Theorem 6.25; Yap [95])

8.* Suppose $G \subseteq K_{n,n}$ is d-regular. Is G Type 1 for any d such that $4 \le d \le n - 2$?

9. Let G be a graph of odd order $2n + 1$ and w be a vertex of minimum degree $\delta(G)$. Suppose G has a clique K_k such that $w \notin V(K_k)$ and $e(G) - \delta(G) > n(\Delta(G) + 1) - k$. Prove that $\chi_T(G) \ge \Delta(G) + 2$.

10. Applying Theorems 6.21 and 6.20, show that except for the two graphs $G(4,1,1,1,1)$ and $G(5,2,1,1,1)$, any graph $G = G(n_5, n_4, ..., n_1)$ of order $2n + 1$ is Type 1.

11. Let $G = K_n + O_r$, the join of K_n and O_r. Prove that G is Type 2 if and only if $n + r$ is even and $\binom{r}{2} < \lceil \frac{n}{2} \rceil$. (Hint: Apply Theorem 6.1; Chen, Fu and Ko [-a])

12. Suppose G is a polar graph $G(K_n, O_r)$. If

$$\Delta(G) \ge \max \{d(v) | v \in V(O_r)\} + n,$$

prove that G is Type 1. (Chen, Fu and Ko [-a])

13. Let G be a polar graph. If $\Delta(G)$ is even, prove that G is Type 1. (Chen, Fu and Ko [-a])

14. Prove that for any odd integer $n \ge 5$, there exists an n-total-colouring φ_1 of K_n having vertex set $\{x_1, x_2, ..., x_n\}$ and an n-total-colouring φ_2 of K_n having vertex set $\{y_1, y_2, ..., y_n\}$ such that $\varphi_1(x_i) = \varphi_2(y_i) = i$, $i = 1, 2, ..., n$ and $\varphi_1(x_1 x_r) = \alpha = \varphi_2(y_n y_r)$, where $2 \le r \le n - 1$. (Yap [93])

15. Let G be a complete bipartite graph having bipartition (X, Y), where $X = \{x_1, x_2, ..., x_n\}$, $Y = \{y_1, ..., y_n\}$ and $n = 2m \ge 2$. Prove that G has an n-edge-colouring φ using n colours $c_1, ..., c_n$ such that

$$\varphi(x_i y_i) = c_n, \quad i = 1, ..., n;$$
$$\varphi(x_{i+1} y_i) = c_i, \quad i = 1, ..., n - 1.$$

(Yap [93]; Note that this result can be used to prove Theorem 6.35 when $|G| = 2n$, n even.)

16. Let G be a graph having vertex set $V = \{1, 2, ..., 2n + 1\}$ and edge set

$$E = (\{(i, j)|1 \le i < j \le 2n - 1\} \setminus \{(1, 2)\}) \cup \{(1, 2n), (2, 2n), ..., (2n, 2n + 1)\}.$$

Prove that G is Type 2. (Contributed by B. L. Chen)

17. Let G be a graph of even order $m \ge 6$ having $\Delta(G) = 2n - 1$. Suppose G is Type 2. Prove that if $e(\tilde{G}) + \alpha'(\tilde{G}) = n - 2$ and $\chi_T(G - e) < \chi_T(G)$ for any edge e of G, then $\tilde{G} = O_{p_1} \cup ... \cup O_{p_r}$ for some odd integers $p_1, p_2, ..., p_r$. (Hamilton, Hilton and Hind [-a])

18. Let G be a graph of odd order $2n + 1 \ge 39$ having $\Delta(G) = 2n - 3$. Suppose $def(G) = 1$. Prove that $\chi_T(G) = \Delta(G) + 2$ and $\chi_T(G - e) < \chi_T(G)$ for any edge e of G if and only if $\bar{G}$ does not contain a K_3. (Hamilton, Hilton and Hind [-a])

19. Let d be an even integer and let G be a d-regular graph of odd order. Suppose G contains an independent set S where $|S| = |G| - d$ and $G - S$ has a perfect matching M. Prove that G is Type 1.

20. Let G be a graph of even order $2n \ge 6$ such that $\bar{G}$ is bipartite and contains a 1-factor. Suppose n is odd. Prove that G is Type 1. (Hilton and Zhao [-a]) (Hint: Apply the proof technique of Case 1 in the proof of Theorem 6.34.)

21. Suppose G is a graph such that the core G_Δ of G contains a perfect matching or near perfect matching $M = \{x_1 y_1, x_2 y_2, ..., x_m y_m\}$ and the other possible edges of G_Δ are as follows: $x_i y_j, x_i x_j$ where $1 \le j < i \le m$; and $x_i z$, $1 \le i \le m$ if $|G_\Delta|$ is odd. Prove that G is Class 1. (Chew and Yap [-a]; for a proof use Yap and Chu [-a]. This result is useful in determining $\chi_T(G)$ for certain graphs G.)

CHAPTER 7

TOTAL CHROMATIC NUMBER OF PLANAR GRAPHS

In this chapter we shall determine the total chromatic number of some planar graphs. The main results are:

(i) For any planar graph G having $\Delta(G) \geq 8$, $\chi_T(G) \leq \Delta(G) + 2$.

(ii) For any planar graph G having $\Delta(G) \geq 14$, $\chi_T(G) = \Delta(G) + 1$. (Borodin [89])

(iii) For any outerplanar graph G having $\Delta(G) \geq 3$, $\chi_T(G) = \Delta(G) + 1$. (Z. F. Zhang, J. X. Zhang and J. F. Wang [88])

The proof of (i) given in this chapter makes use of the Four-Colour Theorem and a very nice result of Vizing which says that if G is a planar graph having $\Delta(G) \geq 8$, then $\chi'(G) = \Delta(G)$. In fact, if Vizing's Planar Graph Conjecture (which says that if G is a planar graph having $\Delta(G) \geq 6$, then $\chi'(G) = \Delta(G)$) is true, then following the proof of (i) we can show that the TCC also holds for planar graphs G having $\Delta(G) \geq 6$. We only present a proof of (iii) for the case $\Delta(G) \geq 4$. The proof of (iii) for the case $\Delta(G) = 3$ is tedious and we leave it as an exercise.

By taking $W = V(G)$ in Lemma 4.2, we have:

Lemma 7.1 For any graph G having $\chi(G) \leq 4$,

$$\chi_T(G) \leq \chi'(G) + 2.$$

We shall now use this lemma to prove Theorem 7.2.

Theorem 7.2 For any planar graph G having $\Delta(G) \geq 8$,

$$\chi_T(G) \leq \Delta(G) + 2.$$

Proof. By the Four-Colour Theorem, $\chi(G) \leq 4$. By a well-known theorem of Vizing which says that if G is a planar graph having $\Delta(G) \geq 8$ then $\chi'(G) = \Delta(G)$ (see Yap [86; p.50]). By Lemma 7.1, we now have $\chi_T(G) \leq \Delta(G) + 2$. $//$

The proof of Theorem 7.2 uses the Four-Colour Theorem. Borodin [89] proved, without using the Four-Colour Theorem, that for any planar graph G,

$$\chi_T(G) \le \Delta(G) + 3 \quad \text{if } \Delta(G) \ge 6;$$
$$\chi_T(G) \le \Delta(G) + 2 \quad \text{if } \Delta(G) \ge 9; \quad \text{and}$$
$$\chi_T(G) = \Delta(G) + 1 \quad \text{if } \Delta(G) \ge 14.$$

In the following we shall prove another lemma and use it to prove that if G is a planar graph having $\Delta(G) \ge 14$, then $\chi_T(G) = \Delta(G) + 1$. We require the following three terms: The weight $w(e)$ of an edge $e = xy$ in a graph G is $d(x) + d(y)$. A 2-alternating cycle $C : v_1 v_2 ... v_m v_1$ of G is a cycle of even length m such that $d(v_1) = d(v_3) = ... = d(v_{m-1}) = 2$. A j-vertex of G is a vertex of degree j in G.

Lemma 7.3 (Borodin [89]) Suppose G is a planar graph having $\delta(G) \ge 2$. Then either G has an edge of weight at most 15 or G contains a 2-alternating cycle.

Proof. Suppose the lemma is false and G is a plane graph which is a counter-example. Then $w(e) \ge 16$ for any $e \in E(G)$ and

(1) for any $x, z \in V(G)$, there is at most one 2-vertex y having $N(y) = \{x, z\}$.

Delete all the 2-vertices from G and embed the resulting graph into a plane triangulation H as follows: For each d-vertex u in G, where $d = 3, 4, 5$, we let $N(u) = \{u_1, u_2, ..., u_d\}$ in clockwise order. We first add edges joining u_i and u_{i+1} (if $u_i u_{i+1} \notin E(G)$) so that $u_1 u_2 ... u_d u_1$ becomes a cycle. We then turn this new graph into a plane triangulation H in any way. Let H' be obtained from H by restoring back all the 2-vertices of G.

Let S be the set of d-vertices in G, where $d = 3, 4, 5$. Let $H^* = H - S$. By the given condition that $w(e) \ge 16$ for any edge e of G and the construction of H, S is independent in H. Since H is a plane triangulation and $w(e) \ge 16$ for any edge e of G, we have

(2) $$d_H(v) \le 2d_{H^*}(v) \quad \text{for any} \quad v \in V(H^*).$$

Since H^* is planar, it contains a vertex of degree at most 5. Suppose H^* contains a vertex v such that $d_{H^*}(v) \le 4$. Then by (2), $d_H(v) \le 8$. Since $d_{H^*}(v) \le 4$, v is adjacent to at least one d-vertex in G. Let u be a d-vertex adjacent to v in G. If v

is not adjacent to any 2-vertex in G, then $w(vu) \leq 8 + 5 = 13$, a contradiction. On the other hand if v is adjacent to at least one 2-vertex, then by (1), it is adjacent to at most $d_{H^*}(v)$ 2-vertices and thus $d_G(v) \leq d_H(v) + 4 \leq 8 + 4 = 12$. Hence G has an edge of weight at most $12 + 2 = 14$, another contradiction.

The above shows that H^* contains a vertex v of degree 5. Let p_i be the number of i-vertices in H^* and p_i^* be the number of i-vertices in H^* which are adjacent to at least one 2-vertex in H'. Let q_2 be the number of edges in H' which are incident with 2-vertices. Now, by (2), $d_{H^*}(v) = 5$ implies that $d_H(v) \leq 10$. Since $w(vu) \geq 16$ for any $u \in V(G)$, and $d_H(v) \leq 10$, v is incident with at least one 2-vertex in H'. Hence $p_5^* = p_5$.

Since p_2 is also the number of 2-vertices in H, using simple counting, we have

$$q_2 = 2p_2 \geq 4p_5^* + 2p_6^*$$

(because $w(e) \geq 16$ in H', each 5-vertex in H^* is incident with at least four 2-vertices in H' and each 6-vertex in H^* is incident with at least two 2-vertices in H').

Finally, since H is a plane triangulation, by Euler's formula, we have

$$p_5^* = p_5 \geq 12 + p_7 + 2p_8 + 3p_9 + \dots \geq p_7^* + p_8^* + p_9^* + \dots$$

Hence $p_2 \geq 2p_5^* + p_6^* \geq p_5^* + p_6^* + p_7^* + p_8^* + \dots$ and $q_2 \geq p_2 + p_5^* + p_6^* + p_7^* + \dots$. Now if H' does not contain any 2-alternating cycle, then all the edges of H' incident with the 2-vertices form paths in H'. Suppose

$$P : u_1 u_2 u_3 \dots$$

is a typical 2-alternating path. Then

$$d_{H'}(u_1) \geq 5, \quad d_{H'}(u_2) = 2, \quad d_{H'}(u_3) \geq 5, \dots .$$

Hence $p_2 + p_5^* + p_6^* + p_7^* + \dots > q_2$, a contradiction to the fact that

$$q_2 \geq p_2 + p_5^* + p_6^* + p_7^* + \dots .$$

Consequently H' contains a 2-alternating cycle C which is also a 2-alternating cycle of G, another contradiction. //

Theorem 7.4 (Borodin [89]) Suppose G is a planar graph having $\Delta(G) \geq 14$. Then

$$\chi_T(G) = \Delta(G) + 1.$$

Proof. We can use induction on $v(G)$ or $e(G)$ to show that this theorem is true if $\delta(G) \leq 1$. Hence we assume that $\delta(G) \geq 2$. By Lemma 7.3, G either has an edge e such that $w(e) \leq 15$ or contains a 2-alternating cycle C. Let $\Delta = \Delta(G)$.

First suppose G has an edge $e = xy$ such that $w(e) \leq 15$. We assume that $d(x) \leq 7$. By induction, $G - e$ has a total-colouring π using $\Delta + 1$ colours $c_1, c_2, ..., c_{\Delta+1}$. Denote $C' = \{c_1, c_2, ..., c_{\Delta+1}\}$, and $\pi[x] = \{\pi(x)\} \cup \{\pi(xz)|z \in N(x), z \neq y\}$. Suppose $\pi(x) = \pi(y)$. We first recolour vertex x by a colour $c \in C' \backslash \{\pi[x] \cup \{\pi(z)|z \in N(x)\}\}$. Hence we can always assume that $\pi(x) \neq \pi(y)$. Next, suppose we have the extreme cases that $\Delta = 14$ and $|\{\pi[x] \cup \{\pi(yw)|w \in N(y)\}\}| = 15$. Then we can recolour vertex x by a colour in $\{\pi(yw)|w \in N(y)\}$ and so we have $|\{\pi[x] \cup \{\pi(yw)|w \in N(y)\}\}| \leq 14$. Hence we can choose $c' \in C' \backslash \{\pi[x] \cup \{\pi(yw)|w \in N(y)\}\}$ to colour the edge e. Thus π can be extended to a total-colouring of G using the same set C' of $\Delta + 1$ colours.

Next suppose G contains a 2-alternating cycle $C : v_1 v_2 ... v_{2t}$, where $d(v_1) = d(v_3) = ... = d(v_{2t-1}) = 2$. By induction $G' = G - \{v_1, v_3, ..., v_{2t-1}\}$ has a total-colouring φ using $\Delta + 1$ colours from $C' = \{c_1, c_2, ..., c_{\Delta+1}\}$ (if $\Delta(G') = \Delta - 1$ we can add one new vertex and join it to a vertex v of degree $\Delta - 1$ in G'). Let $e_1 = v_1 v_2$, $e_2 = v_2 v_3$, ..., $e_{2t} = v_{2t} v_1$. We first extend φ to a partial total-colouring of $G' \cup \{e_1, e_2, ..., e_{2t}\}$ as follows:

If $\varphi[v_2] = \varphi[v_4] = ... = \varphi[v_{2t}]$, then we choose $c, c' \in C' \backslash \varphi[v_2]$ to colour the edges $e_1, e_2, ..., e_{2t}$ so that alternate edges receive the same colour c or c'. Otherwise without loss of generality we assume that $\varphi[v_{2t}] \neq \varphi[v_2]$ and we set $\varphi(e_1) \in \varphi[v_{2t}] \backslash \varphi[v_2]$, $\varphi(e_2) \in C' \backslash \{\varphi[v_2] \cup \{\varphi(e_1)\}\}$, ..., $\varphi(e_{2t}) \in C' \backslash \{\varphi[v_{2t}] \cup \{\varphi(e_{2t-1})\}\}$. It is then clear that we can further extend φ to a total-colouring of G by assigning some suitable colours to the vertices $v_1, v_3, ..., v_{2t-1}$. (there are many choices of colours for each of these vertices).

To prove that the induction works for some initial subgraph of G, we delete more and more edges and ignoring the isolated vertices so that we can eventually end up with a "critical" graph H which has only one vertex v of degree Δ such that v is incident with a pendant vertex. Clearly by deleting the pendant edge e' from H and apply Theorem 7.2 we can see that $\Delta + 1 \leq \chi_T(G) = \chi_T(H - e') \leq (\Delta - 1) + 2 = \Delta + 1$. Hence $\chi_T(G) = \Delta + 1$.   //

To end this section we shall prove (iii) for the case $\Delta(G) \geq 4$. If G has a cut-vertex v and G_1 and G_2 are induced subgraphs of G such that $G = G_1 \cup G_2$, $V(G_1) \cap V(G_2) = \{v\}$, then clearly

$$\chi_T(G) = \max\{\chi_T(G_1), \chi_T(G_2), d(v) + 1\}.$$

Hence to prove that (iii) is true we need only to prove it for the case that G is 2-connected. Since G is outerplanar, we can now assume that all the vertices of G bordering the outer region form a Hamilton cycle $C = v_0 e_0 v_1 e_1 ... v_{n-1} e_{n-1} v_0$.

We shall require the following lemma to prove Theorem 7.6.

Lemma 7.5 Let G be a 2-connected outerplanar graph having $\Delta(G) \geq 4$. Then at least one of the following two conditions holds:

(i) G has a 2-vertex adjacent to a vertex of degree at most 3;

(ii) G has two 2-vertices adjacent to a common vertex of degree 4.

Proof. Assume that G has been embedded in the plane so that all its vertices bordering the outer region form a Hamilton cycle C.

We shall prove this lemma by induction on $|G|$. For $|G| = 3, 4$ or 5, this lemma can be verified easily. Hence we assume that $|G| \geq 6$. Let x be a 2-vertex in G. (It is well-known that any 2-connected outerplanar graph of order at least 3 has at least two 2-vertices) and let $N(x) = \{y, z\}$. We assume that both y and z are of degree at least 4 otherwise there is nothing to prove. Let

$$G' = \begin{cases} G - x & \text{if } yz \in E(G) \\ G - x + yz & \text{if } yz \notin E(G) \end{cases}$$

Then G' a 2-connected outerplanar of order $|G| - 1$ and we can apply the induction hypothesis.

Suppose that G' has two 2-vertices u and v adjacent to a common vertex w of degree 4 in G'. Since y and z are adjacent vertices of the Hamilton cycle C' which forms the border of the outer region of G', $w \neq y, z$. Hence u and v are adjacent to a common vertex w of degree 4 in G.

Suppose that G' has a 2-vertex x' adjacent to a vertex y' of degree at most 3 in G'. If $G' = G - x + yz$, then $y' \neq y, z$ because $d_{G'}(y) \geq 4$ and $d_{G'}(z) \geq 4$. Hence x' is adjacent to a vertex y' of degree at most 3 in G. Thus we can assume that $G' = G - x$. If $y' = y$ or z, say $y' = y$, then x and x' are two 2-vertices in G adjacent to a common vertex y of degree 4 in G. On the other hand if $y' \neq y$ or z, then x' is adjacent to a vertex y' of degree at most 3 in G.   //

Theorem 7.6 (Zhang, Zhang and Wang [88]) For any outerplanar graph G having $\Delta(G) \geq 4$,

$$\chi_T(G) = \Delta(G) + 1.$$

Proof. Clearly we need only to consider the case that G is 2-connected. We shall prove this theorem by induction on $|G|$. The smallest order of a graph G having $\Delta(G) = \Delta$ is $\Delta + 1$. If Δ is even, then $\Delta + 1 \leq \chi_T(G) \leq \chi_T(K_{\Delta+1}) = \Delta + 1$ from which it follows that $\chi_T(G) = \Delta + 1$. If Δ is odd, it is not difficult to show that if G is outerplanar, then $\chi_T(G) = \Delta + 1$ (Exercise 7(3)). Hence we assume that $|G| \geq \Delta + 2$. We also assume that the vertices of G bordering the outer region of the plane form a Hamilton cycle C.

By Lemma 7.5, G either has a 2-vertex x adjacent to a vertex y of degree at most 3 or has two 2-vertices u and v adjacent to a common vertex w of degree 4.

Suppose G has a 2-vertex x adjacent to a vertex y of degree at most 3. Let the other vertex adjacent to x be z. Let

$$G' = \begin{cases} G - x & \text{if } yz \in E(G) \\ G - x + yz & \text{if } yz \notin E(G). \end{cases}$$

Then G' is an outerplanar graph of order $|G| - 1$ having $\Delta - 1 \leq \Delta(G') \leq \Delta$ and by induction G' has a total-colouring φ using $\Delta + 1$ colours from the colour set $S = \{1, 2, ..., \Delta + 1\}$. We shall modify φ to yield a total-colouring of G using the same set S of colours.

Since $\Delta \geq 4$, there is at least one colour $\alpha \in S$ missing at vertex y. We now set $\varphi(yx) = \alpha$, $\varphi(zx) = \gamma$ where $\gamma = \varphi(yz)$, and $\varphi(x) \in S \setminus \{\alpha, \gamma, \varphi(y), \varphi(z)\}$ to obtain a required $(\Delta + 1)$-total-colouring of G.

Finally we settle the case that G has two 2-vertices u and v adjacent to a common vertex w of degree 4. Let the other vertex adjacent to u be w'. We have already proved the theorem if $d(w') \leq 3$. Hence we assume that $d(w') \geq 4$.

Suppose $ww' \notin E(G)$. Let $G' = G - u + ww'$. Then again G' satisfies either (i) or (ii) of Lemma 7.5. The graph G' is called a __contraction__ of G at the 2-vertex u. If we contract all this kind of 2-vertices one-by-one we will eventually obtain a graph in which all the two neighbours of each 2-vertex in G are of degree at least 4 and are adjacent in G. This means that we can assume, without loss of generality, that the two neighbours w and w' of u we are considering now are adjacent in G.

Let w'' be the other vertex adjacent to v in G. By symmetry we can assume that $ww'' \in E(G)$ and $d(w'') \geq 4$. Let $G' = G - u - v$. Then by induction G' has a total colouring φ using $\Delta + 1$ colours from S. Let α and β be the colours missing at w' and w'' respectively. Let α' and β' be two colours missing at w.

Suppose first that $\alpha \neq \beta$. Then we can assume that $\alpha' \neq \alpha$ and $\beta' \neq \beta$. We now modify φ to a total-colouring of G using the same set S of colours by setting $\varphi(uw') = \alpha$, $\varphi(vw'') = \beta$, $\varphi(uw) = \alpha'$, $\varphi(vw) = \beta'$, $\varphi(u) \in S \setminus \{\varphi(w'), \alpha, \alpha', \varphi(w)\}$ and $\varphi(v) \in S \setminus \{\varphi(w''), \beta, \beta', \varphi(w)\}$.

Suppose next that $\alpha = \beta$. Clearly if $\alpha \notin \{\alpha', \beta'\}$, then we can apply the previous method to extend φ a total-colouring of G using the same set S of colours. Suppose $\alpha \in \{\alpha', \beta'\}$. Let $\alpha = \alpha'$ and $\gamma = \varphi(w)$. We now recolour the vertex w by colour α. Then colours β' and γ are missing at vertex w and we can apply the previous method to modify φ to a required total-colouring of G. //

Exercise 7

1. Prove that for any graph G such that $\chi'(G) \leq 8$,

$$\chi_T(G) \leq \chi'(G) + 4.$$

2. Verify Lemma 7.5 for the cases $|G| = 3, 4$ and 5.

3. Suppose $\Delta \geq 3$ is an odd integer. Prove that if G is an outerplanar graph of order $\Delta + 1$ and having $\Delta(G) = \Delta$, then $\chi_T(G) = \Delta + 1$.

4. Prove that if G is an outerplanar graph having $\Delta(G) = 3$, then $\chi_T(G) = 4$.
    (Z. F. Zhang, J. X. Zhang and J. F. Wang [88])

5. Let G be a planar graph with $\delta(G) \geq 3$. Prove that

   (i) G has an edge e with $w(e) \leq 13$;

   (ii) either G has an edge e with $w(e) \leq 11$ or G contains a 3-altering 4-cycle.
   (Borodin [89])

6. Applying Lemma 7.3 and Exercise 7(5), prove that if G is a planar graph with $\Delta(G) \leq 8$, then

$$\chi_T(G) \leq \Delta(G) + 3.$$

(Borodin [89])

7.* Determine the smallest possible integer Δ such that all the planar graphs G with $\Delta(G) \geq \Delta$ are Type 1. (Borodin [89]. Remarks: Theorem 7.4 shows that $\Delta \leq 14$.)

CHAPTER 8

SOME UPPER BOUNDS FOR THE TOTAL CHROMATIC NUMBER OF GRAPHS

Since the Total Colouring Conjecture is still unsolved, one may look for some other upper bounds for $\chi_T(G)$. The first paper which touched on this subject is Kostochka [77] where it is proved that for most multigraphs (with a few unsettled cases) G with $\Delta(G) \geq 6$, $\chi_T(G) \leq \frac{3}{2}\Delta(G)$. In this section we present some upper bounds for $\chi_T(G)$ due to Chetwynd, Häggkvist, Hind, McDiarmid, and Sánchez-Arroyo.

(i) $\chi_T(G) \leq \chi'(G) + \lfloor \frac{1}{3}\chi(G) \rfloor + 2$. (Sánchez-Arroyo [95])

(ii) For any graph G and any positive integer $k \leq \Delta(G)$,

$$\chi_T(G) \leq \chi'(G) + \left\lceil \frac{\chi(G)}{k} \right\rceil + k.$$

(Hind [90]. Remarks: From (ii) one can deduce that for any graph G, $\chi_T(G) \leq \chi'(G) + 2\lceil \sqrt{\chi(G)} \rceil$. Note that this bound is better than the bound given in (i) for large $\chi(G)$.)

(iii) If G is a graph of order n and k is an integer such that $k! \geq n$, then

$$\chi_T(G) \leq \chi'(G) + k.$$

(Chetwynd and Häggkvist [92]. Note that when $v(G)$ is large and $\Delta(G)$ is large with respect to $v(G)$, this result superseeds (ii).)

(iv) For any graph G,

$$\chi_T(G) \leq \Delta(G) + 2\left\lceil \frac{v(G)}{\Delta(G)} \right\rceil + 1.$$

(Hind [92])

Sánchéze-Arroyo [95] uses Lemma 8.1 to prove Theorem 8.2 which improves an earlier result of McDiarmid and Sánchez-Arroyo [93].

Lemma 8.1 Suppose $G = (V, E)$ is a grpah having $\Delta(G) \leq 2$. Let V_1, V_2 and V_3 be disjoint independent sets of vertices of G and let $W = V_1 \cup V_2 \cup V_3$. Then there exists a partial total-colouring $\psi : E \rightarrow \{1, 2, 3, \alpha\}$ such that

(i) if $x \in V_i$, then $\psi(x) = i$, $i = 1, 2, 3$;

(ii) if $xy \in E$, and $\phi(xy) = \alpha$, then x, $y \in W$.

Proof. The proof is essentially the same as the proof of Lemma 4.1.   //

Theorem 8.2 (Sánchez-Arroyo [95]) For any graph G,

$$\chi_T(G) \leq \chi'(G) + \lfloor \frac{1}{3}\chi(G) \rfloor + 2.$$

Proof. By Brooks' theorem, except when G is a complete graph or an odd cycle, $\chi(G) \leq \Delta(G)$.

Since we have proved that the TCC holds for K_n and odd cycles, we need only to consider the case

$$p = \chi(G) \leq \Delta(G) \leq \chi'(G) = m.$$

Let ϕ be a p-vertex-colouring and ψ be an m-edge-colouring of G using colours $1, 2, ..., p$ and $c_1, c_2, ..., c_m$ respectively. Let $V_1, ..., V_p$ and $E_1, ..., E_m$ be the colour classes of ϕ and ψ respectively.

We consider three cases.

Case 1. $p = 3k$.

In this case for each $j = 1, 2, ..., k$, let

$$W_j = V_{3(j-1)+1} \cup V_{3(j-1)+2} \cup V_{3j} \quad \text{and} \quad E'_j = E_{3(j-1)+1} \cup E_{3(j-1)+2}.$$

We now apply Lemma 8.1 to the graph $H_j = (V, E'_j)$ with the subset $W_j \subseteq V$ to obtain a partial total-colouring ϕ_j of $W_j \cup E'_j$ using colours from the set $\{3(j-1) + 1, 3(j-1) + 2, 3j, \alpha\}$. It is clear that the union of these colourings $\phi_1, ..., \phi_k$ form a partial total-colouring of $V \cup (E'_1 \cup ... \cup E'_k)$ and consequently G has a total-colouring using

$$(m - p) + p + \frac{1}{3}p + 1 = \chi_e(G) + \lfloor \frac{1}{3}\chi(G) \rfloor + 1$$

colours $c_{m-p+1}, ..., c_m, 1, 2, ..., p, c_3, c_6, ..., c_k$ and α.

Case 2. $p = 3k + 1$.

In this case $p - 1 = 3k$ and we argue as Case 1 with the condition that each vertex in V_p is recoloured with colour α. The number of colours used to achieve a total-colouring for G is

$$(m - (p - 1)) + (p - 1) + \frac{1}{3}(p - 1) + 1 = \chi_e(G) + \lfloor \frac{1}{3}\chi(G) \rfloor + 1.$$

Case 3. $p = 3k + 2$.

In this case $p - 2 = 3k$ and we argue as Case 2. The number of colours used to achive a total-colouring for G is

$$(m - (p - 2)) + (p - 2) + \frac{1}{3}(p - 2) + 2 = \chi'(G) + \lfloor \frac{1}{3}\chi(G) \rfloor + 2. \quad //$$

We shall require the following lemma to prove Theorem 8.4.

Lemma 8.3 (Hind [90]) Let H be a graph and W an independent set of vertices of H. Suppose

$$\phi : W \to \{1, 2, ..., \chi'(H)\}$$

is a colouring. Then ϕ can be extended to a proper colouring of $W \cup E(H)$ using the set of colours $\{1, 2, ..., \chi'(H)\} \cup \{\alpha\}$ such that if $\phi(e) = \alpha$ then one of the endvertices of e is in W.

Proof. We prove this result by induction on $k = \chi'(H)$. If $k = 1$, the result is clearly true. Now suppose $k \geq 2$ and $E(H) = E_1 \cup ... \cup E_k$, where $E_1, ..., E_k$ are disjoint matchings in H. Let $H' = H - E_k$ and $W' = \{w \in W : \phi(w) \neq k\}$. Since $\chi'(H') = k - 1$, by induction, we can extend ϕ to a proper colouring of $W' \cup E(H')$ using the set of colours $\{1, 2, ..., k - 1\} \cup \{\alpha\}$ such that if $\phi(e) = \alpha$ then one of the endvertices of e is in W'.

We next define a colouring ψ of $W \cup E(H)$ by setting

$$\psi|_W = \phi, \quad \psi|_{E(H')} = \phi \quad \text{and} \quad \psi(e) = k \quad \text{for any } e \in E_k.$$

If ψ is not a proper colouring, then there exists a vertex-edge pair (x_0, x_0x_1) such that $x_0 \in W \backslash W'$, $\phi(x_0) = k$ and $x_0x_1 \in E_k$. Let $P = x_0x_1x_2...x_r$ be the maximal

(k, α)-path (the path where the edges $x_0 x_1, x_1 x_2, \ldots$ are alternately coloured with colours k and α and there is no vertex other than x_{r-1} such that $\phi(x_{r-1} x_r) = k$ or α).

Since W is independent in H and if $\phi(e) = \alpha$ then one of the endvertices of e is in W', we have

$$x_{2i} \in W', \quad x_{2i+1} \notin W, \quad \text{and} \quad \psi(x_{2i-1} x_{2i}) = \alpha \quad \text{for all} \quad i \leq \left\lfloor \frac{r}{2} \right\rfloor.$$

Interchanging the colours k and α in P, we reduce the number of vertices $x_0 \in W \setminus W'$ such that $\phi(x_0) = k$ and x_0 is incident with an edge coloured k. As the (k, α)-paths having initial vertex in $W \setminus W'$ are vertex-disjoint, we may repeat the same process for all these (k, α)-paths at the same time and we obtain a required proper colouring ψ of $W \cup E(H)$. //

Theorem 8.4 (Hind [90]) For any grpah G and any positive integer $k \leq \Delta(G)$,

$$\chi_T(G) \leq \chi'(G) + \left\lceil \frac{\chi(G)}{k} \right\rceil + k.$$

Proof. Let $\Delta = \Delta(G)$. Since the TCC holds for odd cycles and complete graphs, we may assume that $G \neq C_n$ or K_n. Hence, by Brooks' theorem, $\chi = \chi(G) \leq \Delta$. Let ϕ be a fixed (proper) vertex-colouring of G using colours $1, 2, \ldots, \chi$ and let $V_1, V_2, \ldots, V_\chi$ be the colour classes of ϕ.

Let $q = \chi'(G)$. Let θ be a (proper) edge-colouring of G using the set of colours $\{1, 2, \ldots, \chi, \chi + 1, \ldots, q\}$ and let $E_1, E_2, \ldots, E_q$ be the colour classes of θ.

For k satisfying $1 \leq k \leq \chi$, let $m = \lceil \frac{\chi}{k} \rceil$ and let $I_1 = \{1, 2, \ldots, m\}$, $I_2 = \{m + 1, m + 2, \ldots, 2m\}, \ldots, I_{k-1} = \{(k-2)m + 1, (k-2)m + 2, \ldots, (k-1)m\}$, $I_k = \{(k-1)m + 1, (k-1)m + 2, \ldots, \chi\}$. For each $i = 1, 2, \ldots, k$, let G_i be the spanning subgraph of G having edge-set $E_{(i-1)m+1} \cup \ldots \cup E_{im}$. By definition, G_i is m-edge-colourable. Let $W_i = \{v \in V(G) | \phi(v) \in I_i\}$, and $F(G_i) = \{xy \in E(G_i) | x, y \in W_i\}$. Now the spanning subgraph G'_i of G having edge-set $E(G_i) \setminus F(G_i)$ satisfies the conditions of Lemma 8.3 and so $\phi|_{W_i}$ can be extended to a proper colouring η_i of $W_i \cup E(G'_i)$ using the set of colours $I_i \cup \{\alpha_i\}$.

Next it is clear that $G[F(G_1) \cup \ldots \cup F(G_k)]$ has a proper edge-colouring ψ using m colours $\beta_1, \ldots, \beta_m$ because each $F(G_i)$ is m-edge-colourable and an edge in $F(G_i)$ is not adjacent to an edge in $F(G_j)$, $j \neq i$. Consequently we can combine $\eta_1, ..., \eta_k$ and ψ to get a total-colouring η of $G[E_1 \cup ... \cup E_\chi]$ using the set of colours $\{1, 2, ..., \chi\} \cup \{\alpha_1, ..., \alpha_k\} \cup \{\beta_1, ..., \beta_m\}$, where $\eta|_{V(G)} = \phi$.

Finally η can be extended to a $(q + m + k)$-total-colouring of G by restoring back the set of coloured edges $E_{\chi+1} \cup ... \cup E_q$ because ϕ is a proper vertex-colouring of G and thus for any edge $e = xy$ in G, $\phi(x) \neq \phi(y)$.   //

Corollary 8.5 For any graph G,

$$\chi_T(G) \leq \chi'(G) + 2\lceil \sqrt{\chi(G)} \rceil.$$

Proof. Taking $k = \lceil \sqrt{\chi(G)} \rceil$ in Theorem 8.4, the result follows.   //

The following theorem was proved independently by Chetwynd and Häggkvist [92] as well as by McDiarmid and Reed [93].   //

Theorem 8.6 (Chetwynd and Häggkvist [92]; McDiarmid and Reed [93]) If G is a graph of order n and k is an integer such that $k! \geq n$, then

$$\chi_T(G) \leq \chi'(G) + k.$$

Proof. We can assume tht G is connected and $G \neq C_n$ or K_n. Let $q = \chi'(G)$. By Brooks' theorem, G has a vertex-colouring φ using q colours $1, 2, ..., q$ and so G has a collection $\mathcal{S} = \{S_1, S_2, ..., S_q\}$ of disjoint independent sets of vertices where each $v \in S_i$ has $\varphi(v) = i$. Since $\chi'(G) = q$, G has also a collection $\mathcal{M} = \{M_1, M_2, ..., M_q\}$ of disjoint matchings. Since the TCC holds for small n and very large vertex degree, we can let $2 \leq k \leq q - 2 (\leq n - 3)$.

For a given bijection π from $\mathcal{M}$ to $\mathcal{S}$, let the 'violation graph' G' be the subgraph of G having

$$E(G') = \{xy \in E(G) | xy \in M \Rightarrow x \in \pi(M) \text{ or } y \in \pi(M)\}.$$

Clearly if we colour the edges in M with colour i, where $\pi(M) = S_i$ and recolour the edges of the 'violation graph' G' with $\chi'(G')$ new colours, then we have a (proper) total-colouring of G. Hence

$$\chi_T(G) \leq q + \chi'(G') \leq q + 1 + \Delta(G').$$

To complete the proof we shall show that there exists a bijection π from $\mathcal{M}$ to $\mathcal{S}$ for which $\Delta(G') \leq k - 1$.

Consider a vertex v in G having neighbourhood W with $|W| \geq k$. Let $\mathcal{C}$ be the collection of all subsets $W' \subseteq W$ with $|W'| = k$ such that $\varphi(x) \neq \varphi(y)$ for any $x, y \in W'$. Also, for each $W' \in \mathcal{C}$, let $A(W')$ be the event that, for each $w \in W'$, the matching $M_i \in \mathcal{M}$ containing the edge vw is mapped to $S_j \in \mathcal{S}$, where $w \in S_j$ or $v \in S_j$. Then

$$P\{A(W')\} = \frac{1}{q(q-1)...(q-k+1)} = \frac{(q-k)!}{q!}$$

where $P\{A(W')\}$ is the probability that the event $A(W')$ occurs.

Since $|\mathcal{C}| \leq \binom{|W|}{k}$, we now have

$$P\{d_{G'}(v) \geq k\} \leq \binom{|W|}{k} \frac{(q-k)!}{q!}.$$

Since the neighbours of v receive colours different from $\varphi(v)$, $|W| \leq \Delta - 1$ where $\Delta = \Delta(G)$. Hence $\binom{|W|}{k} < \binom{\Delta}{k}$. Also $q = \Delta$ or $\Delta + 1$ implies that $\frac{(q-k)!}{q!} \leq \frac{(\Delta-k)!}{\Delta!}$. Hence

$$P\{d_{G'}(v) \geq k\} < \binom{\Delta}{k} \frac{(\Delta-k)!}{\Delta!} = \frac{1}{k!}$$

and thus

$$P(\Delta(G') \geq k) < \frac{n}{k!} \leq 1.$$

Consequently there exists a bijection $\pi : \mathcal{M} \to \mathcal{S}$ such that $\Delta(G') \leq k - 1$ as required.
//

We shall next apply Theorem 5.11 to prove Theorem 8.7.

Theorem 8.7 (Hind [92]) For any graph G,

$$\chi_T(G) \leq \Delta(G) + 2\left\lceil \frac{v(G)}{\Delta(G)} \right\rceil + 1.$$

Proof. Let $V = V(G)$, $\Delta = \Delta(G)$ and $k = \lceil \frac{|V|}{\Delta} \rceil$. Theorem 5.11 guarantees the existence of a vertex-colouring φ of G using colours $1, 2, ..., \Delta + 2k + 1$ such that every colour class contains at most k vertices.

Let E^* be an edge subset of G, containing the maximum number of edges, for which there is a proper total-colouring $\psi : V \cup E^* \to \{1, 2, ..., \Delta + 2k + 1\}$ with $\psi|_V = \varphi$.

Suppose $E^* \neq E(G)$. Let $w_0 x_0 \in E(G) \setminus E^*$ and let $H = G[E^* \cup \{w_0 x_0\}]$, the induced subgraph of G having edge set $E^* \cup \{w_0 x_0\}$. In the following we shall show that ψ can be modified to a total-colouring ψ^* of $G[E^* \cup \{w_0 x_0\}]$ such that $\psi^*|_V = \varphi$. This contradicts the choice of E^*. Consequently G is $(\Delta + 2k + 1)$-total-colourable.

We first define an associate mapping

$$\bar{\psi} : V \to \mathbb{P}(\{1, 2, ..., \Delta + 2k + 1\}),$$

where $\mathbb{P}(\{1, 2, ..., \Delta + 2k + 1\})$ is the power set of $\{1, 2, ..., \Delta + 2k + 1\}$, by letting $\bar{\psi}(v)$ to be the set of colours in $\{1, 2, ..., \Delta + 2k + 1\}$ which are not assigned to ψ to v nor to an edge incident with v. Clearly $|\bar{\psi}(w_0)| \geq 2k + 1$, $|\bar{\psi}(x_0)| \geq 2k + 1$ and $|\bar{\psi}(v)| \geq 2k$ for any other vertex v of G.

We next define a sequence of pairwise disjoint subsets of $N_G(w_0)$ as follows: We let $X_0 = \{x_0\}$, $X_1 = \{x \in N_G(w_0) | \psi(w_0 x) \in \bar{\psi}(x_0)\} \setminus X_0$, i.e. the set of vertices $x \neq x_0$ in $N_G(w_0)$ such that $\psi(w_0 x) \in \bar{\psi}(x_0)$ and in general for $i = 1, 2, ...,$ let

$$X_i = \{x \in N_G(w_0) | \psi(w_0 x) \in \bar{\psi}(x') \quad \text{for some } x' \in X_{i-1}\} \setminus (\bigcup_{j=0}^{i-1} X_j).$$

Since $N_G(w_0)$ is finite and the sets $X_0, X_1, X_2, ...$ are pairwise disjoint, there exists a smallest integer $m \geq 1$ for which $X_m = \phi$. From the definition of X_i, it follows that $X_{m+1} = X_{m+2} = ... = \phi$. Let $X = X_0 \cup X_1 \cup ... \cup X_{m-1}$.

If $x \in X$ and $r(x)$ denotes the index for which $x \in X_{r(x)}$, then there exists a sequence of vertices $v_0(x) = x$, $v_1(x), ..., v_{r(x)}(x) = x_0$ such that

(i) $v_i(x) \in X_{r(x)-i}$ for $i = 0, 1, ..., r(x)$; and

(ii) $\psi(w_0 v_i(x)) \in \bar{\psi}(v_{i+1}(x))$ for $i = 0, 1, ..., r(x) - 1$.

Obviously $r(x_0) = 0$ and the sequence for x_0 is the trivial sequence x_0. For each $x \in X$ we shall assume that a single (fixed) "nested" sequence of vertices

$$S(x) : v_0(x) = x, \ v_1(x), ..., v_{r(x)}(x) = x_0$$

satisfying (i) and (ii) has been chosen.

We shall use the following observation later:

(O.) If $x \in X$ and $\gamma \in \bar{\psi}(x) \setminus \{\psi(w_0)\}$, then there exists $y \in X$ such that $\psi(w_0 y) = \gamma$.

Suppose otherwise. We first note that if there exists $y \in N_G(w_0) \setminus X$ such that $\psi(w_0 y) = \gamma$, then by definition, $y \in X_{r(x)+1}$ and thus $y \in X$, a contradiction. Now let

$$S(x) : v_0(x) = x, \ v_1(x), ..., v_{r(x)}(x) = x_0$$

be a "nested" sequence of vertices constructed above. We can modify ψ to a total-colouring ψ' of H by letting

$$\psi'(w_0 v_i(x)) = \psi(w_0 v_{i-1}(x)) \quad \text{for } i = 1, 2, ..., r(x)$$

and $\psi'(w_0 v_0(x)) = \gamma$, and letting ψ' fix the other elements of H. The existence of such a total-colouring of H contradicts the choice of E^*. This proves observation (O.)

We next define a set of colours B: For each $i = 0, 1, 2, ..., m - 1$, let

$$B_i = \bigcup_{x \in X_i} \bar{\psi}(x) \setminus \{\psi(w_0)\} \quad \text{and let} \quad B = B_1 \cup B_2 \cup ... \cup B_{m-1}.$$

Clearly from the definitions of X and B we know that if $x \in X$ then $\bar{\psi}(x) \setminus \{\psi(w_0)\} \subseteq B$. Furthermore, if $x \neq x_0$, then $\psi(w_0 x) \in B$. We define a bipartite graph J having bipartition (X, B) such that for any $x \in X$ and $\beta \in B$, $x\beta \in E(J)$ if and only if $\beta \in \bar{\psi}(x) \setminus \{\psi(w_0)\}$ or $\psi(w_0 x) = \beta$. Since $|\bar{\psi}(x)| \geq 2k$ for each $x \in X$ and $|\bar{\psi}(x_0)| \geq 2k + 1$, it follows that

$$d_J(x) \geq 2k \quad \text{for every} \quad x \in X.$$

Also, by observation (O.), for each $\beta \in B$ there exists a unique $x \in X \setminus \{x_0\}$ such that $\psi(w_0 x) = \beta$. Thus $|X| = |B| + 1$ and this implies that there exists $\beta^* \in B$ such that $d_J(\beta^*) \geq 2k + 1$.

Let α be a fixed colour in $\bar{\psi}(w_0)$. By observation (O.), for any $x \in X$, $\alpha \notin \bar{\psi}(x) \setminus \{\psi(w_0)\}$. Hence for each $x \in X$, either $\psi(x) = \alpha$ or $\psi(xy) = \alpha$ for some edge $xy \in E^*$. In particular, this is true for any $x \in N_J(\beta^*)$. If $\psi(x) = \alpha$, then $\beta^* \in \bar{\psi}(x) \setminus \{\psi(w_0)\}$ implies that x is a trivial (α, β^*)-path in G. If $\psi(xy) = \alpha$ for some edge $xy \in E^*$, then the (α, β^*)-path in G containing x has initial vertex either

x or w_0 (in case $\psi(w_0 x) = \beta^*$) and we say that this (α, β^*)-path starts at x. If an (α, β^*)-path P contains two distinct vertices $x_1, x_2 \in N_J(\beta^*)$, then at least one of x_1 and x_2 is a terminus and thus the endvertices of P are not coloured with α or β^*. Hence we can interchange the colours α and β^* in P to get another total-colouring of $G[E^*]$. (In this case we say that P is "swoppable".) On the other hand if each $x \in N_J(\beta^*)$ is contained in exactly one (α, β^*)-path, then since $|N_J(\beta^*)| \geq 2k + 1$, there are at least $2k + 1$ such paths. However, since each vertex colour class of ψ contains at most k vertices, there are at most $2k$ of the endvertices of these paths coloured with either α or β^* by ψ and thus at least one of such paths is swoppable.

Consider the vertices x in $N_J(\beta^*)$ which have a swoppable (α, β^*) path starting at x. Let x^* be such a vetex which has a shortest sequence $S(x^*)$ and let P be a swoppable (α, β^*)-path starting at x^*. Let t be the termius of P.

We consider three cases separately:

Case 1. $\psi(w_0 x^*) = \beta^*$.

The choice of x^* as having $r(x^*)$ a minimum ensures that $v_1(x^*) \neq t$. We now interchange the colours α and β^* in P and recolour the edges $w_0 v_i(x^*)$ with $\psi(w_0 v_{i-1}(x^*))$ for $i = 1, 2, ..., r(x^*)$. This yields a (proper) total-colouring of H, a contradiction.

Case 2. $\psi(w_0 x^*) \neq \beta^*$ and $t \neq w_0$.

This case can be settled in a similar way as Case 1.

Case 3. $\psi(w_0 x^*) \neq \beta^*$ and $t = w_0$.

Let $y^* \in V(G)$ be such that $\psi(w_0 y^*) = \beta^*$. We consider two subcases separately:

Subcase 3(i). If $x^* \notin V(S(y^*))$, then interchange the colours α and β^* in P and recolour the edges $w_0 v_i(y^*)$ with $\psi(w_0 v_{i-1}(y^*))$ for $i = 1, 2, ..., r(y^*)$. This yields a (proper) total-colouring of H, a contradiction.

Subcase 3(ii). Suppose $x^* \in V(S(y^*))$. If $x^* \neq v_1(y^*)$, then we can settle this case in a similar way as Subcase 3(i). Hence we assume that $x^* = v_1(y^*)$. Now since $d_J(\beta^*) \geq 2k + 1 \geq 3$, $N_J(\beta^*)$ contains another vertex z^*. Suppose $y^* \notin S(z^*)$. Then interchanging colours α and β^* in $P : x^* y^* w_0$, recolouring $w_0 z^*$ with β^* and recolouring the edges $w_0 v_i(z^*)$ with $\psi(w_0 v_{i-1}(z^*))$ for $i = 1, 2, ..., r(z^*)$, we obtain a (proper) total-colouring of H, a contradiction. On the other hand, suppose $y^* \in S(z^*)$. Let $y^* = v_j(z^*)$. Then $x^* = v_{j+1}(z^*)$. In this case we first recolour $w_0 z^*$ with β^*, recolour the edges $w_0 v_i(z^*)$ with $\psi(w_0 v_{i-1}(z^*))$ for $i = 1, 2, ..., j$, and then recolour $y^* x^*$ with β^*, $w_0 x^*$ with α, and the edges $w_0 v_i(x^*)$ with $\psi(w_0 v_{i-1}(x^*))$ for $i = 1, 2, ..., r(x^*)$. Now once again after interchanging the colours assigned to the edges of P we will obtain a (proper) total-colouring of H, a contradiction. Hence G is $(\Delta + 2k + 1)$-total-colourable.   //

Exercise 8

1. Prove that for any graph G,

$$\chi_T(G) \le \chi'(G) + \left\lfloor \frac{\chi(G)}{2} \right\rfloor + 1.$$

(Kostochka [77b]; Sánchez-Arroyo [91])

2. Let H be a graph obtained by inserting a new vertex into an edge of $K_{\Delta,\Delta}$, $\Delta \ge 6$. Let G be the graph obtained by identifying the three minor vertices of three copies of H. Then the identified vertex is a cut-vertex (the only cut-vertex) of G. Colour the bipartitions of each copy of $K_{\Delta,\Delta}$ with colours 1 and 2, and colour the cut-vertex of G with colour 3. Prove that this vertex-colouring ψ of G cannot be extended to a total-colouring of G using $\Delta + 2$ colours $1, 2, 3, ..., \Delta + 2$. (This problem is contributed by H. R. Hind.)

3. Prove that the graph G constructed in Problem 2 above is $(\Delta+2)$-total-colourable. (Remarks: Problems 2 and 3 show that if G is such that $\chi_T(G) \le \Delta(G)+2$ then it is not true that any vertex-colouring of G using at most $\Delta(G) + 2$ colours can be extended to a total-colouring of G using the same set of $\Delta + 2$ colours.)

4.* Does there exist a graph G having $\Delta(G) = 3$ and G has a (proper) vertex-colouring φ using exactly five colours such that φ cannot be extended to a total-colouring of G using the same set of five colours? (This question is raised by H. R. Hind.)

CHAPTER 9

CONCLUDING REMARKS

To end this book we mention in this last chapter some other important and interesting results concerning total-colourings of graphs which we are unable to discuss in detail because lack of time and space.

1. Total chromatic number of most graphs. Let p_n be the probability of graphs of order n having $\chi_T(G) \geq \Delta(G) + 2$. McDiarmid [90] proves that $p_n \to 0$ as $n \to \infty$. McDiarmid and Reed [93] prove that $p_n \leq n^{-\left(\frac{1}{8} + o(1)\right)n}$ as $n \to \infty$ and that the corresponding probability of graphs having $\chi_T(G) > \Delta(G) + 2$ is $o(c^{n^2})$ for some $0 < c < 1$. Hind [-b] also proves that almost every graph satisfies the TCC. These results provide very strong evidence for the truth of the TCC.

2. Computational complexity of determining the total chromatic number. Sánchez-Arroyo [89] proves that determining the total chromatic number is NP-hard. McDiarmid and Sánchez-Arroyo [-a] strengthen this result by showing that even determining the total chromatic number of regular bipartite graphs is NP-hard.

3. Total graphs. The total graph $T(G)$ of a graph G is a graph with vertex set $V(G) \cup E(G)$ and in which two vertices u and v of $T(G)$ are adjacent if and only if they are either two adjacent or two incident elements of G. Clearly $\chi_T(G) = \chi(T(G))$. Since in general it is easier to determine $\chi_T(G)$ by directly colouring the vertices and edges of G rather than to determine $\chi(T(G))$ by colouring the vertices of $T(G)$, we do not present any results on properties of $T(G)$. For more informations on properties of $T(G)$, such as the connectivity, planarity, and characterization etc, see Behzad [67], [69], [71a], Behzad and Chartrand [67], and Behzad and Radjavi [69].

4. Total-chromatic critical graphs. The notion of chromatic index critical graphs are extremely useful in the classification of graphs according to the chromatic index (for details, see Yap [86;p.21]). But, the parallel notion with respect to total chromatic number has not yet proved to have the same importance.

Behzad [71] proved that for any edge e of a graph G, $\chi_T(G - e) \geq \chi_T(G) - 1$. Hence for any edge e of G, either $\chi_T(G - e) = \chi_T(G)$ or $\chi_T(G - e) = \chi_T(G) - 1$. If G has an edge e_1 such that $\chi_T(G - e_1) = \chi_T(G)$, then we remove e_1 from G. Let $G_1 = G - e_1$. If G_1 has an edge e_2 such that $\chi_T(G_1 - e_2) = \chi_T(G_1)(= \chi_T(G))$, then we remove e_2 from G_1. We continue this process until all the edges of G_i (for some i) become "critical", then we obtain a total-chromatic critical graph.

Zhang Zhongfu seems to be the first person to explore further (in an unpublished paper) the critical concept in total-colourings of graphs. However, his definition of criticality is very restricted. He calls a graph G such that $\chi_T(G) > \Delta(G) + 1$ an "edge-critical total-colouring graph" if for any edge e of G, $\chi_T(G-e) = \Delta(G-e)+1$. He then proves several results on this type of critical graphs. Most of his results are actually also true for the general definition of total-chromatic critical graphs which we define below. The proofs of these results are not difficult and we leave them as exercises at the end of this chapter.

A connected graph G is said to be <u>total-chromatic critical</u> if $\chi_T(G - e) < \chi_T(G)$ for any edge e of G. The following are some total-chromatic critical graphs:

(i) The cycles C_n, where n is not a multiple of 3 are "critical" (Exercise 1(1)).

(ii) Let G be a graph of order $2n$ and having $\Delta(G) = 2n-1$. If $e(\tilde{G})+\alpha'(\tilde{G}) = n-1$, then G is total-chromatic critical (Theorem 6.1).

(iii) If G is a graph of order $2n$ and having $\Delta(G) = 2n - 2$ then G is total-chromatic critical if and only if $\bar{G} = S_{2n-3} \cup S_1$ (Theorem 6.9).

(iv) If $J \subseteq K_{n,n}$ and $e(J) + \alpha'(J) = n - 1$, then $K_{n,n} - E(J)$ is total-chromatic critical (Corollary 6.26).

(v) Suppose G is a graph of order $2n + 1$ and having $\Delta(G) = 2n - 1$. Let w be a vertex of minimum degree in G. Then G is total-chromatic critical if and only if

$$e(\bar{G} - w) + \alpha'(G - w) = n$$

(Theorem 6.24).

(vi) Suppose G is a graph of even order and having $\Delta(G) = |G| - 3 \geq 3$. If $\bar{G}$ is the union of two odd cycles, then G is total-chromatic critical (Theorem 6.36).

(vii) Suppose G is a graph of odd order $2n + 1$ and G is $(2n - 2)$-regular. Then G is total-chromatic critical if $\bar{G} \not\supseteq K_3$ (see Remarks (ii), page 93).

Hamilton and Hilton [91] have constructed all total-chromatic critical graphs G of order at most 16 and having $\Delta(G) = 3$. Hamilton, Hilton and Hind [-a] give a list of all graphs of order at most ten which are total-chromatic critical. In this paper, they also rephrase Hilton's Conformability Conjecture (see p.13) in terms of total-chromatic critical graphs as follows:

Hilton's Conjecture: Let G be a graph having $\Delta(G) \geq \frac{1}{2}(v(G) + 1)$. Then G is total-chromatic critical if and only if either G is non-conformable and G contains no proper non-conformable subgraph having maximum degree $\Delta(G)$, or $\Delta(G)$ is even and G is obtained by inserting a new vertex into an edge of $K_{\Delta(G)+1}$.

Hamilton, Hilton and Hind [-a] also establish the Conformability Conjecture in the case when $v(G)$ is odd, $\Delta(G) \geq \frac{1}{3}\{\sqrt{7}v(G) + \mathrm{def}(G) + 1\}$ and $\mathrm{def}(G) \leq v(G) - \Delta(G) - 1$. Furthermore, in the same paper, they also show that if the Conformability Conjecture is true, then all total-chromatic critical graphs of G satisfying $\frac{1}{2}v(G)+1 \leq \Delta(G) \leq \frac{3}{4}v(G) - 1$ are in fact regular.

5. <u>Total chromatic number of infinite graphs.</u> Behzad and Radjavi [76] proved that if the TCC is true for (finite) graphs, then it is true for infinite graphs of bounded degree.

6. <u>Coupled colourings of plane graphs.</u> The following "mixed chromatic numbers" have some similarities to the total chromatic number of plane graphs. Let G be a plane graph. Denote by $\chi_{ef}(G)$ the minimum number of colours needed to colour the edges and faces of G such that two elements from $E(G) \cup F(G)$ ($F(G)$ is the set of faces of G) are adjacent or incident receive distinct colours. The notation $\chi_{vf}(G)$ and $\chi_{vef}(G)$ are defined similarly. Borodin [87] has proved, among other results, that

(i) if $\Delta(G) \geq 11$, then $\chi_{ef}(G) \leq \Delta(G) + 3$;

(ii) if $\Delta(G) \geq 12$, then $\chi_{vef}(G) \leq \Delta(G) + 4$;

(iii) if $\Delta(G) \geq 17$, then $\chi_{ef}(G) \leq \Delta(G) + 1$;

(iv) if $\Delta(G) \geq 18$, then $\chi_{vef}(G) \leq \Delta(G) + 2$ provides that G does not contain self-adjacent faces.

(These results are taken from the Mathematical Review : 89d : 05069 reviewed by Ioan Tomescu. The interested readers may have to find the definition of self-adjacent faces from Borodin's original paper.)

Borodin [90b], has proved, among other results that

(v) for any planar G having no separating 3-cycles,

$$\chi_{ef}(G) \leq \Delta(G) + 4.$$

(This result is taken from Mathematical Review : 92e : 05035 reviewed by Joseph Zaks. For the definition of separating 3-cycles, the interested readers may have to refer to Borodin's paper [90b].)

Borodin [84,85] proved the following conjecture (quoted from B. Toft [-a]):

Ringel's Conjecture (1968). For any plane graph G, $\chi_{vf}(G) \leq 6$.

Finally we mention a longstanding conjecture:

Kronk-Mitchem Conjecture (1973). For any plane graph G,

$$\Delta(G) + 1 \leq \chi_{vef}(G) \leq \Delta(G) + 4.$$

(Note that Kronk and Mitchem proved that their conjecture is true if $\Delta(G) = 3$. Also Zhang Zhongfu, Wang Jianfang and Wang Wei Feng have proved that if G is a 2-connected outerplanar graph G having $\Delta(G) \geq 7$, then $\chi_{vef}(G) = \Delta(G) + 1$. We have also learnt that O. V. Borodin had proved that Kronk-Mitchem conjecture is true if $\Delta(G) \geq 12$ and also that if $\Delta(G) \geq 18$, then $\chi_{vef}(G) \leq \Delta(G) + 2$.)

7. **A generalization of total-colouring.** Hind [-b] extends the concept of a generalized edge-colouring introduced by Hakimi and Kariv (see S. L. Hakimi and O. Kariv, A generalization of edge-coloring in graphs, Jour. Graph Theory 10 (1986), 139-154) to that of a generalized total-colouring:

Let $C = \{c_1, c_2, ..., \}$ be a set of colours. Let G be a graph and $m : V(G) \rightarrow \mathbb{N} = \{1, 2, ...\}$ be a function. Then a (proper) generalized-colouring of G with respect to m is a mapping

$$\pi : S \rightarrow C,$$

where $S \subseteq V(G) \cup E(G)$, such that for each $c_i \in C$ and each $v \in V(G)$,

$$|\{vx \in E(G) | \pi(vx) = c_i\}| \begin{cases} \leq m(x) & \text{if } \pi(x) \neq c_i \\ < m(x) & \text{if } \pi(v) = c_i \end{cases}$$

and

$$|\{x \in N(v) | \pi(x) = c_i\}| < m(v) \quad \text{if } \pi(v) = c_i.$$

If $S = E(G)$ then π is a generalized edge-colouring as considered by Hakimi and Kariv. If $S = V(G)$ then π is called a generalized vertex-colouring of G and if $S = V(G) \cup E(G)$ then π is called a generalized total-colouring of G. (Note that $m(v) = 1$ for all $v \in V(G)$ yields the usual edge-colouring, vertex-colouring, and total-colouring of G, respectively.)

The generalized total chromatic number of G with respect to m, denote by $\chi''_m(G)$, is the smallest positive integer k for which there exists a generalized total-colouring π of G such that

$$\pi : VE(G) \rightarrow \{1, 2, ..., k\}.$$

Hind [-b] proves that for any graph G,

$$\chi''_m(G) \leq \max_{v \in V(G)} \left\{ \left\lceil \frac{d(v) + 1}{m(v)} \right\rceil + 2 \left\lceil \sqrt{\frac{d(v) + 1}{m(v)}} \right\rceil \right\}$$

which generalizes Theorem 8.6.

8. Total chromatic number and density of a graph. The density $t(G)$ of a graph G is define by

$$t(G) = \max \left\{ \frac{e(H)}{v(H)} \mid H \text{ is a nontrivial subgraph of } G \right\}.$$

Chen and Wu [94] prove the following results:

(I) if G is a graph with $t(G) \leq 2$, then $\chi_T(G) \leq \Delta(G) + 2$. (As a corollary if G is planar having girth at least 4, then $\chi_T(G) \leq \Delta(G) + 2$.)

(II) if G is a connected graph with $t(G) \leq 1 + \frac{k}{3}$ ($k = 1, 2, 3$) and $\Delta(G) \geq 3 + \frac{(k^2+1)}{2}$, then G is Type 1. (As a corollary if G is a planar graph, having maximum degree Δ and girth g then G is Type 1 provided one of the following holds:

(i) $\Delta \geq 8$ and $g \geq 4$

(ii) $\Delta \geq 5$ and $g \geq 5$

(iii) $\Delta \geq 4$ and $g \geq 8$.

9. <u>Number of different total-colourings of a graph.</u> Two total-colourings φ and π of a graph G using the same set of $\chi_T(G)$ colours are said to be the <u>same</u> if π can be obtained from φ by permutations of colours and applying automorphisms of G. It can be verified that K_3 and K_5 have only one total-colouring but K_7 has two distinct total-colourings. Up to now there are no papers published on this subject.

10. <u>A relationship between an Edge-Colouring Conjecture and a Total-Colouring Conjecture.</u> Chetwynd and Hilton posed the following.

Edge-Colouring Conjecture. Suppose G is a graph having $\Delta(G) > \frac{1}{3}|G|$. Then G is Class 2 if and only if it contains no overfull subgraph H with $\Delta(H) = \Delta(G)$.

(A graph G is <u>overfull</u> if $e(G) > \Delta(G) \lfloor \frac{1}{2}|G| \rfloor$.)

Hilton and Zhao [91] prove that if the Edge-Colouring Conjecture is true for graphs of even order having $\Delta(G) > \frac{1}{2}|G|$, then Hilton's conjecture (see p.13) is true for graphs of odd order having

$$\delta(G) \geq \frac{3}{4}|G| - \frac{1}{4} \quad \text{and} \quad \det(G) \geq 2(\Delta(G) - \delta(G) + 1).$$

(Remarks: The condition that $\delta(G) \geq \frac{3}{4}|G| - \frac{1}{4}$ can be replaced by $\Delta(G) \geq \frac{3}{4}|G| - \frac{1}{4}$.)

Exercise 9

1. Suppose G is a total-chromatic critical graph and G is not a star. Prove that $\chi_T(G) \geq \Delta(G) + 2$.

2. Suppose G is a total-chromatic critical graph and G is not a star. Prove that G does not contain a cut-edge.

3. Let G be a total-chromatic critical graph having $\Delta(G) \geq 2$. Suppose G contains a vertex u of degree 2 and $N(u) = \{v, w\}$. Prove that $d(v) = d(w) = \Delta(G)$. (This problem is contributed by Zhang Zhongfu.)

4. Suppose G is a total-chromatic critical graph which contains a vertex of degree 2. Prove that $\chi_T(G) = \Delta(G) + 2$. (This problem is contributed by Zhang Zhongfu.)

5. Suppose G is a total-chromatic critical graph and $G \neq K_2$. Prove that $\chi_T(G - e) = \chi_T(G) - 1$.

6. Let G be a 2-connected outerplanar graph. Suppose $\Delta(G) = 3$. Prove that $4 \leq \chi_{ef}(G) \leq 5$. (Hu Guangzhang and Zhang Zhongfu [-a])

7. Let G be a 2-connected outerplanar graph. Suppose $4 \leq \Delta(G) \leq 6$. Prove that $\chi_{ef}(G) \leq 7$. (Hu Guanzhang and Zhang Zhongfu [-a])

8. Let G be a 2-connected outerplanar graph. Suppose $\Delta(G) \geq 7$. Prove that $\chi_{ef}(G) = \Delta(G)$. (Hu Guanzhang and Zhang Zhongfu [-a])

9. Which of the following two graphs is total-chromatic critical? (Hamilton and Hilton [91])

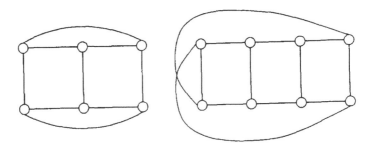

Figure 9.1

10. Let G be a graph of even order $2n \geq 6$ and having $\Delta(G) = 2n - 1$. Prove that G is total-chromatic critical if and only if either $e(\bar{G}) + \alpha'(\bar{G}) = n - 1$, or $e(\bar{G}) + \alpha'(\bar{G}) = n - 2$ and $\bar{G}$ consists of a number of disjoint complete graphs of odd order. (Hamilton, Hilton and Hind [-a])

11. Give an example showing that if H is a proper subgraph of a graph G, it is not necessary true that $\chi_{ef}(H) \leq \chi_{ef}(G)$. (This problem is contributed by Zhang Zhongfu.)

12. Give an example showing that if H is a proper subgraph of a graph G, it is not necessary true that $\chi_{vef}(H) \leq \chi_{vef}(G)$. (This problem is contributed by Zhang Zhongfu.)

13.* Let G be a total-chromatic critical graph. Is it true that the number of major vertices of G must be at least $\frac{1}{2}v(G)$? (Hamilton, Hilton and Hind [-a])

REFERENCES

L. D. Andersen and A. J. W. Hilton, Thanks Evans!, *Proc. London Math. Soc.* (3) 47 (1983), 507-522.

M. Behzad, *Graphs and Their Chromatic Numbers*, Doctoral Thesis Michigan State University (1965).

_____, A criterion for the planarity of the total graph of a graph, *Proc. Cambridge Philos. Soc.* 63 (1967), 679-681.

_____, The connectivity of total graphs, *Bull. Austral. Math. Soc.* 1 (1969), 175-181.

_____, A characterization of total graphs, *Proc. Amer. Math. Soc.* 26 (1971), 383-389.

_____, The total chromatic number of a graph : A survey. In *Combinatorial Mathematics and its Applications* (Ed. D. J. A. Welsh), Acad. Press (1971), 1-9.

_____, Total concepts in graph theory, *Ars Combin.* 23 (1987), 35-40.

M. Behzad and G. Chartrand, An introduction to total graphs, in : *Theory of Graphs* (International Symposium, Rome 1966), Gordon and Breach (1967).

M. Behzad, G. Chartrand and J. K. Cooper Jr., The colour numbers of complete graphs, *J. London Math. Soc.* 42 (1967), 226-228.

M. Behzad, G. Chartrand and L. Lesniak-Foster, *Graphs and Digraphs*, Wadsworth International (1979).

M. Behzad and H. Radjavi, Structure of regular total graphs, *J. London Math. Soc.* 44 (1969), 433-436.

_____, Chromatic numbers of infinite graphs, *J. Combin. Theory* 21 (1976), 195-200.

J. C. Bermond, Nombre chromatique total du graphe r-parti complet, *J. London Math. Soc.* (2) 9 (1974), 279-285.

B. Bollobás and A. J. Harris, List colourings of graphs, *Graphs and Combinatorics* 1 (1985), 115-127.

O. V. Borodin, Solution of Ringel's problem about vertex bound colouring of planar graphs and colouring of 1-planar graphs (in Russian), *Metody Diskret Analiz* 41 (1984), 12-26.

————, Consistent colouring of 1-embedded graphs (in Russian), in: Graphen und Netzwerke-Theorie und Anwendungen 30 Intern. Wiss. Koll. TH I'lmenau (1985), 19-20.

————, Coupled colorings of graphs on a plane (Russian), *Metody Diskret. Analiz, Novosibirsk*, 45 (1987), 21-27; 95.

————, On the total colouring of planar graphs, *J. reine angew Math.* 394 (1989), 180-185.

————, An extension of Kotzig's theorem and edge-list colouring of planar graphs (Russian), *Matem. Zametki* 48, No. 6 (1990), 22-28.

————, A structural property of planar graphs and the simultaneous colouring of their edges and faces, *Math. Slovaca* 40, No. 2 (1990), 113-116.

————, Simultaneous coloring of edges and faces of plane graphs, *Discrete Math.* 128 No. 1-3 (1994), 21-33.

B. L. Chen and H. L. Fu, Total colourings of graphs of order $2n$ having maximum degree $2n - 2$, *Graphs and Combinatorics* 8 (No. 2) (1992), 119-123.

B. L. Chen, H. L. Fu and M. T. Ko, Total chromatic number and chromatic index of split graphs. (manuscript)

Chen Dong-Ling and Wu Jian-Liang, Total colouring of some graphs, in *"Combinatorics, Graph Theory, Algorithms and Applications"*, Proc. of the Third China-USA International Conference on Graph Theory, Beijing 1993 (Eds. Y. Alavi, D. R. Lick and Liu Jiuqiang) World Scientific (1994), 17-20.

A. G. Chetwynd, Total colourings, in : *Graph Colourings* (Eds. R. Nelson and R. J. Wilson), Pitman Research Notes 218 (1990), 65-77.

A. G. Chetwynd and R. Häggkrist, Some upper bounds on the total and list chromatic numbers of multigraphs, *J. Graph Theory* 16 (1992), 503-516.

A. G. Chetwynd and A. J. W. Hilton, The chromatic index of graphs of even order with many edges, *J. Graph Theory* 8 (1984), 463-470.

————, Some refinements of the total chromatic number conjecture, Nineteenth Southeastern Conf. on Combinatorics, *Congr. Numer.* 66 (1988), 195-215.

————, 1-factorizing regular graphs of high degree - an improved bound, *Discrete Math.* 75 (1989), 103-112.

————, A Δ-subgraph condition for a graph to be Class 1, *J. Combin. Theory, Ser. B*, 46 (1989), 37-45.

123

――――, The chromatic index of graphs with large maximum degree, where the number of vertices of maximum degree is relatively small, *J. Combin. Theory, Ser. B*, 48 (1990), 45-66.

A. G. Chetwynd, A. J. W. Hilton and Zhao Cheng, On the total chromatic number of graphs of high minimum degree, *J. London Math. Soc.* (2) 44 (1991), 193-202.

K. H. Chew and H. P. Yap, Chromatic index and total chromatic number of graphs of high degree, Research Report No. 425, June 1990, Department of Mathematics, National University of Singapore.

――――, Total chromatic number of complete r-partite graphs, *J. Graph Theory*, 16 (No. 6) (1992), 629-634.

R. J. Cook, Complementary graphs and total chromatic numbers, *SIAM J. Appl. Math.* 27 (4) (1974), 626-628.

J. K. Dugdale and A. J. W. Hilton, The total chromatic number of regular graphs of order $2n$ and degree $2n - 3$, *J. Combin. Inform. System Sci.* 15, No. 14 (1990), 103-110.

――――, The total chromatic number of regular graphs whose complement is bipartite, *Discrete Math.* 126 (1994), 87-98.

S. Fiorini and R. J. Wilson, *Edge-Colouring of Graphs*, Research Notes in Mathematics, Vol. 16, Pitman, London (1977).

E. Flandrin, H. A. Jung and H. Li, Hamiltonism, degree sums and neighborhood conditions, *Discrete Math.*, 90 (1991), 41-52.

R. Häggkvist and A. Chetwynd, Some upper bounds on the total and list chromatic numbers of multigraphs, *J. Graph Theory* 16 (1992), 503-516.

A. Hajnal and E. Szemerédi, Proof of a conjecture of P. Erdös, In : *Combinatorial Theory and its Applications II* (Ed. P. Erdös, A. Rényi and V. T. Sós), Colloq. Math. Soc. J. Bolyai 4 (1970), 601-623.

G. M. Hamilton and A. J. W. Hilton, Graphs of maximum degree 3 and order at most 16 which are critical with respect to the total chromatic number, *J. Combin. Math. Comput.*, 10 (1991), 129-149.

G. M. Hamilton, A. J. W. Hilton and H. R. Hind, Graphs which are critical with respect to the total chromatic number. (preprint)

A. J. W. Hilton, Recent results on edge-colouring graphs, with applications to the total-chromatic number, *J. of Combin. Math. and Combin. Computing*, 3 (1988), 121-134.

_____, A total-chromatic number analogue of Plantholt's theorem, *Discrete Math.*, 79 (1989/90), 169-175.

_____, The total chromatic number of nearly complete bipartite graphs, *J. Combin. Theory, Ser. B*, 52 (No. 1) (1991), 9-19.

_____, Recent results on the total chromatic number, *Discrete Math.*, 111(1993), 323-331.

A. J. W. Hilton and H. R. Hind, Total chromatic number of graphs having large maximum degree, *Discrete Math.* 117 (1993), 127-140.

A. J. W. Hilton and Zhao Cheng, The relationship between an edge-colouring conjecture and a total-colouring conjecture, *J. Combin. Math. Combin. Comput.* 10 (1991), 83-95.

_____, A sufficient condition for a graph *G* to be Type 1. (preprint)

H. R. Hind, An upper bound for the total chromatic number, *Graphs and Combinatorics*, 6 (1990), 153-159.

_____, *Restricted Edge-colourings*, Ph.D. Thesis, Cambridge University (1988).

_____, An upper bound for the total chromatic number of dense graphs, *J. Graph Theory* 16 (1992), 192-203.

_____, A generalization of total colourings. (preprint)

_____, Almost every graph satisfies the Total Colouring Conjecture. (preprint)

D. G. Hoffman and C. A. Rodger, The chromatic index of complete multipartite graphs, *J. Graph Theory* 16 (1992), 159-164.

_____, The total chromatic number of complete multipartite graphs. (preprint)

Hu Guangzhang and Zhang Zhongfu, On the edge-face total colourings of plane graphs, *J. of Tsinghua Univ.*, Vol. 32, No. 3 (1992), 18-23.

Kathryn F. Jones and Jennifer Ryan, Total colouring a graph with maximum degree four, *Ars Combinatoria*, 31 (1991), 277-285.

D. König, *Theorie der endlichen und unendlichen Graphen*, Leipzig (1936). (Reprinted, New York, Chelsea (1950); An English translation is published by Birkäuser, Boston (1990))

A. V. Kostochka, The total coloring of a multigraph with maximal degree 4, *Discrete Math.* 17 (1977), 161-163.

_____, An analogue of Shannon's estimate for total colouring (Russian), *Diskret. Analiz* 30 (1977), 13-22.

_____, Exact upper bound for total chromatic number of a graph, *Proc. 24th Intern. Wiss Kell. Th. Ilmenau* (1979), 33-36.

_____, The total chromatic number of any multigraph with maximum degree 5 is at most seven. (submitted)

H. Kronk and J. Mitchem, A seven-color theorem on the sphere, *Discrete Math.* 5 (1973), 253-260.

C. J. H. McDiarmid, Colourings of random graphs, in : *Graph Colourings* (Eds. R. Nelson and R. J. Wilson), Pitman Research Notes in Mathematics 218 (1990).

C. J. H. McDiarmid and B. Reed, On total colourings of graphs, *J. Combin. Theory, Ser. B,* 57, No. 1 (1993), 122-130.

C. J. H. McDiarmid and A. Sánchez-Arroyo, Total colouring regular bipartite graphs is NP-hard, *Discrete Math.* 124, No. 1-3 (1994), 155-162.

_____, An upper bound for total colouring of graphs, *Discrete Math.* 111 (1993), 389-392.

J. C. Meyer, Nombre chromatique total du joint d'un ensemble stable par un cycle, *Discrete Math.* 15 (1976), 41-54.

J. Ninčák, Algorithm of colouring of regular complete r-partite graphs, *Proc. 24th Intern. Wiss Kell. Th. Ilmenau* (1979), 37-40.

G. Ringel, A six-colour problem on the sphere, In *"Theory of Graphs"* (Eds. P. Erdös and G. Katona), Academic Press, New York (1968), 265-269.

M. Rosenfeld, On the total colouring of certain graphs, *Israel J. Math.* 9 (3) (1971), 396-402.

A. Sánchez-Arroyo, Determining the total colouring number is NP-hard. *Discrete Math.* 78 (1989), 315-319.

_____, *Colourings, Complexity, and Some Related Topics*, Ph.D. Thesis, Linacre College, University of Oxford (1991).

M. A. Seound, Total chromatic numbers, *Appl. Math. Lett.* 5, No. 6 (1992), 37-39.

Shen Minggang, Classification of complete r-partite graphs according to their total chromatic numbers. (preprint)

A. Soifer, A six-coloring of the plane, *J. Combin. Theory, Ser. A*, 61, No. 2 (1992), 292-294.

B. Toft, *Graph Colouring Problems*, Part 1. (unpublished)

N. Vijayaditya, On total chromatic number of a graph, *J. London Math. Soc.* (2), 3 (1971), 405-408.

V. G. Vizing, On an estimate of the chromatic clas of a p-graph (Russian), *Diskret. Analiz* 3 (1964), 25-30.

——, Critical graphs with a given chromatic class (Russian), *Diskret. Analiz* 5 (1965), 9-17.

——, Some unsolved problems in graph theory, *Uspehi Mat. Nauk* 23 (1968), 117-134 = *Russian Math. Surveys* 23 (1968), 125-142.

Wang Chin-Chung, *Total Colourings of Graphs of High Degree*, MSc Thesis, National Chiao Tung University, Taiwan (1990).

Wang Jianfang and Zhang Zhongfu, The total chromatic numbers of a graph and its complemental graph, *Chinese Quarterly J. of Math.* 2 (1987), 44-51.

Wang Weifan, The edge face entire chromatic number of planar graphs with low degree, *Appl. Math. Ser. A* (Chinese Ed.) 8, No. 3 (1993), 300-307.

Xu Baogang, The entire colouring of a maximal planar graph and its homeomorphism graph, *J. Shandong Univ. Nat. Sci. Ed.* 29, No. 1, (1994), 8-12.

Yao Huineng, A simple proof of the outerplanar's total chromatic number theorem, *J. Math. Res. Expo.* 13, No. 3 (1993), 470-472.

H. P. Yap, *Some Topics in Graph Theory*, Cambridge University Press (1986).

——, Total colourings of graphs, *Bull. London Math. Soc.* 21 (1989), 159-163.

——, Generalization of two results of Hilton on total colourings of a graph, *Discrete Math.* 140 (1995), 245-252.

——, Total chromatic number of graphs G having maximum degree $|G| - 3$, In: *"Combinatorics and Graph Theory"* (Eds. H. P. Yap, T. H. Ku, E. K. Lloyd and Z. M. Wang), World Scientific, Singapore (1993), 192-207.

H. P. Yap, B. L. Chen and H. L. Fu, Total chromatic number of graphs of order $2n+1$ having maximum degree $2n - 1$, *J. London Math. Soc.*, (1995), to appear.

H. P. Yap and K. H. Chew, Total chromatic number of graphs of high degree, II, *J. Australian Math. Soc., Ser A*, 53 (1992), 219-228.

H. P. Yap and W. Y. Chu, Total chromatic number of graphs of high degree, *Proc. of the Spring School and International Conf. on Combinatorics and Graph Theory*, World Scientific, Singapore. (to appear)

H. P. Yap and Qizhang Liu, Edge colouring of K_{2n} with spanning star-forests receiving distinct colours. (submitted)

H. P. Yap, Wang Jian-Fang and Zhang Zhongfu, Total chromatic number of graphs of high degree, *J. Australian Math. Soc.*, 47 (Ser. A) (1989), 445-452.

Zhang Xiandi, On total colouring of the join of K_k and $\bar{K}_{n-k}$, *Bull. Institute of Combin. and its Applicatons*, Vol. 7 (1993), 67-72.

Zhang Zhongfu, Liu Yunpei and Hu Guangzhang, On colourings of planar graphs : A survey (preprint).

Zhang Zhongfu and Sun Liang, On the n-total chromatic number of graphs, *Chin. Ann. Math., Ser. A*, 13, No. 1 (1992), 70-75.

Zhang Zhongfu, Zhang Jianxue and Wang Jianfang, The total chromatic number of some graphs, *Scientia Sinica, Ser. A*, (1988), 1434-1441.

Zhang Zhongfu, Zhang Kemin and Li Xiaodong, A property of total coloring of graphs, *J. Nanjing Univ. Math. Bio.* 8, No. 2 (1991), 99-102.

INDEX OF SUBJECTS

INDEX OF NOTATION

$\delta(G)$, minimum vertex degree   1

$\delta_{m,n}$ Kronecker delta function   16

$\Delta(G)$, maximum vertex degree   1

$\pi|_{V(G)}$, restriction of π on $V(G)$ 4

$\pi|_{E(G)}$, restriction of π on $E(G)$ 4

$\chi(G)$, chromatic number of G 3

$\chi'(G)$, chromatic index of G 3

$\chi_T(G)$, total chromatic number of G 4

$\mu(G)$, multiplicity of G 3

$\mu(x,y)$, multiplicity of xy 3

Lecture Notes in Mathematics

For information about Vols. 1–1439
please contact your bookseller or Springer-Verlag

Vol. 1479: S. Bloch, I. Dolgachev, W. Fulton (Eds.), Algebraic Geometry. Proceedings, 1989. VII, 300 pages. 1991.

Vol. 1480: F. Dumortier, R. Roussarie, J. Sotomayor, H. Żołądek, Bifurcations of Planar Vector Fields: Nilpotent Singularities and Abelian Integrals. VIII, 226 pages. 1991.

Vol. 1481: D. Ferus, U. Pinkall, U. Simon, B. Wegner (Eds.), Global Differential Geometry and Global Analysis. Proceedings, 1991. VIII, 283 pages. 1991.

Vol. 1482: J. Chabrowski, The Dirichlet Problem with L²-Boundary Data for Elliptic Linear Equations. VI, 173 pages. 1991.

Vol. 1483: E. Reithmeier, Periodic Solutions of Nonlinear Dynamical Systems. VI, 171 pages. 1991.

Vol. 1484: H. Delfs, Homology of Locally Semialgebraic Spaces. IX, 136 pages. 1991.

Vol. 1485: J. Azéma, P. A. Meyer, M. Yor (Eds.), Séminaire de Probabilités XXV. VIII, 440 pages. 1991.

Vol. 1486: L. Arnold, H. Crauel, J.-P. Eckmann (Eds.), Lyapunov Exponents. Proceedings, 1990. VIII, 365 pages. 1991.

Vol. 1487: E. Freitag, Singular Modular Forms and Theta Relations. VI, 172 pages. 1991.

Vol. 1488: A. Carboni, M. C. Pedicchio, G. Rosolini (Eds.), Category Theory. Proceedings, 1990. VII, 494 pages. 1991.

Vol. 1489: A. Mielke, Hamiltonian and Lagrangian Flows on Center Manifolds. X, 140 pages. 1991.

Vol. 1490: K. Metsch, Linear Spaces with Few Lines. XIII, 196 pages. 1991.

Vol. 1491: E. Lluis-Puebla, J.-L. Loday, H. Gillet, C. Soulé, V. Snaith, Higher Algebraic K-Theory: an overview. IX, 164 pages. 1992.

Vol. 1492: K. R. Wicks, Fractals and Hyperspaces. VIII, 168 pages. 1991.

Vol. 1493: E. Benoît (Ed.), Dynamic Bifurcations. Proceedings, Luminy 1990. VII, 219 pages. 1991.

Vol. 1494: M.-T. Cheng, X.-W. Zhou, D.-G. Deng (Eds.), Harmonic Analysis. Proceedings, 1988. IX, 226 pages. 1991.

Vol. 1495: J. M. Bony, G. Grubb, L. Hörmander, H. Komatsu, J. Sjöstrand, Microlocal Analysis and Applications. Montecatini Terme, 1989. Editors: L. Cattabriga, L. Rodino. VII, 349 pages. 1991.

Vol. 1496: C. Foias, B. Francis, J. W. Helton, H. Kwakernaak, J. B. Pearson, H∞-Control Theory. Como, 1990. Editors: E. Mosca, L. Pandolfi. VII, 336 pages. 1991.

Vol. 1497: G. T. Herman, A. K. Louis, F. Natterer (Eds.), Mathematical Methods in Tomography. Proceedings 1990. X, 268 pages. 1991.

Vol. 1498: R. Lang, Spectral Theory of Random Schrödinger Operators. X, 125 pages. 1991.

Vol. 1499: K. Taira, Boundary Value Problems and Markov Processes. IX, 132 pages. 1991.

Vol. 1500: J.-P. Serre, Lie Algebras and Lie Groups. VII, 168 pages. 1992.

Vol. 1501: A. De Masi, E. Presutti, Mathematical Methods for Hydrodynamic Limits. IX, 196 pages. 1991.

Vol. 1502: C. Simpson, Asymptotic Behavior of Monodromy. V, 139 pages. 1991.

Vol. 1503: S. Shokranian, The Selberg-Arthur Trace Formula (Lectures by J. Arthur). VII, 97 pages. 1991.

Vol. 1504: J. Cheeger, M. Gromov, C. Okonek, P. Pansu, Geometric Topology: Recent Developments. Editors: P. de Bartolomeis, F. Tricerri. VII, 197 pages. 1991.

Vol. 1505: K. Kajitani, T. Nishitani, The Hyperbolic Cauchy Problem. VII, 168 pages. 1991.

Vol. 1506: A. Buium, Differential Algebraic Groups of Finite Dimension. XV, 145 pages. 1992.

Vol. 1507: K. Hulek, T. Peternell, M. Schneider, F.-O. Schreyer (Eds.), Complex Algebraic Varieties. Proceedings, 1990. VII, 179 pages. 1992.

Vol. 1508: M. Vuorinen (Ed.), Quasiconformal Space Mappings. A Collection of Surveys 1960-1990. IX, 148 pages. 1992.

Vol. 1509: J. Aguadé, M. Castellet, F. R. Cohen (Eds.), Algebraic Topology - Homotopy and Group Cohomology. Proceedings, 1990. X, 330 pages. 1992.

Vol. 1510: P. P. Kulish (Ed.). Quantum Groups. Proceedings, 1990. XII, 398 pages. 1992.

Vol. 1511: B. S. Yadav, D. Singh (Eds.), Functional Analysis and Operator Theory. Proceedings, 1990. VIII, 223 pages. 1992.

Vol. 1512: L. M. Adleman, M.-D. A. Huang, Primality Testing and Abelian Varieties Over Finite Fields. VII, 142 pages. 1992.

Vol. 1513: L. S. Block, W. A. Coppel, Dynamics in One Dimension. VIII, 249 pages. 1992.

Vol. 1514: U. Krengel, K. Richter, V. Warstat (Eds.), Ergodic Theory and Related Topics III. Proceedings, 1990. VIII, 236 pages. 1992.

Vol. 1515: E. Ballico, F. Catanese, C. Ciliberto (Eds.), Classification of Irregular Varieties. Proceedings, 1990. VII, 149 pages. 1992.

Vol. 1516: R. A. Lorentz, Multivariate Birkhoff Interpolation. IX, 192 pages. 1992.

Vol. 1517: K. Keimel, W. Roth, Ordered Cones and Approximation. VI, 134 pages. 1992.

Vol. 1518: H. Stichtenoth, M. A. Tsfasman (Eds.), Coding Theory and Algebraic Geometry. Proceedings, 1991. VIII, 223 pages. 1992.

Vol. 1519: M. W. Short, The Primitive Soluble Permutation Groups of Degree less than 256. IX, 145 pages. 1992.

Vol. 1520: Yu. G. Borisovich, Yu. E. Gliklikh (Eds.), Global Analysis – Studies and Applications V. VII, 284 pages. 1992.

Vol. 1521: S. Busenberg, B. Forte, H. K. Kuiken, Mathematical Modelling of Industrial Process. Bari, 1990. Editors: V. Capasso, A. Fasano. VII, 162 pages. 1992.

Vol. 1522: J.-M. Delort, F. B. I. Transformation. VII, 101 pages. 1992.

Vol. 1523: W. Xue, Rings with Morita Duality. X, 168 pages. 1992.

Vol. 1524: M. Coste, L. Mahé, M.-F. Roy (Eds.), Real Algebraic Geometry. Proceedings, 1991. VIII, 418 pages. 1992.

Vol. 1525: C. Casacuberta, M. Castellet (Eds.), Mathematical Research Today and Tomorrow. VII, 112 pages. 1992.